P9-BYA-887

UNHAPPY LANDINGS

WHY AIRPLANES CRASH

THOMAS W. WATSON

HARBOR CITY PRESS, INC.
MELBOURNE, FLORIDA

For information write or call:

Harbor City Press, Inc.,
P.O. Box 361033,
Melbourne, Fl 32936-1033
(407) 255-0908
1-800-741-4890

FIRST EDITION SEPTEMBER 1992

- -

LIBRARY OF CONGRESS CATALOGING-IN-PUBLICATION DATA

Watson, Thomas W. (Thomas Waller). 1922-

Unhappy landings / by Thomas W. Watson. -- 1st ed.

p. cm.

Includes index.

ISBN 0-9630891-0-2

1. Aeronautics--Accidents. 2. Aeronautics--Safety measures. 3. Aeronautics--Accident investigation. 4. United States. National Transportation Safety Board. I. Title.

TL553.5.W37 1992

363.12'4'0973--dc20 92-16913

CIP

- -

PRINTED IN THE UNITED STATES OF AMERICA

COVER DESIGN AND TYPE COMPOSITION BY

BETTE R. JONES

DEDICATION

This book is lovingly dedicated to Dorothy, my wife; our three offspring, Thomas, William, and Cynthia; and to the Glory of Almighty God.

AUTHOR'S COMMENT

The opinions and conclusions expressed herein are those of the author. They represent an analysis of my observations and experiences garnered during a 24-year career as a pilot in the U. S. Marine Corps, and a 20-year tenure as an aircraft accident investigator for our government. A vast majority of the probable causes for accidents under discussion are in consonance with determinations made by the presidentially appointed membership of the National Transportation Safety Board. In those instances where the causes expressed are not in full agreement with the Safety Board's findings and determinations, they represent the author's conclusions based on the factual evidence gathered during the course of the investigations.

In the interest of privacy, efforts have been made to avoid identifying individuals involved in the acidents. Likewise, it has been attempted to refrain from naming the precise locations of various accidents under discussion. However, instances were encountered where it became necessary to violate those precepts.

ACKNOWLEDGEMENTS

Recognition is hereby made, and grateful appreciation is expressed, for the invaluable editorial and technical assistance rendered by a number of individuals during the three years required to complete this book.

Technical editors were Colonel Arthur O. Schmagel, a retired Marine Corps Aviator and consultant to PAN AM; Lieutenant Commander Arnold E. Holstine, a retired Naval Aviator and Safety Board accident investigator; Arthur E. Neumann, a former Naval Aviator and retired Safety Board Supervisory Air Safety Investigator; Lieutenant Colonel E. Douglas Dreifus, a retired Marine Corps Aviator and Safety Board Accident Investigator; cousin Irvin B. Baldwin, President, Commonwealth Jet Service Inc., Richmond, Virginia; and many others, who provided answers to my numerous technical queries. Their wholehearted support ensured that nothing substantive was lost during the rephrasing of rather technical aviation terms into language that should be readily comprehended by the nonaviation-oriented reader.

Editors from outside the aviation community were my wife, Dorothy, who also provided much encouragement; sister, Charlotte Giles, who provided needed insight from the lay reader's perspective; brother Cecil C. Watson, a retired business executive who continually urged that I write these accounts just as I told them; brother, Dr. Richard G. Watson, an Ordained Presbyterian Minister and Vice President for Academic Affairs at Reformed Theological Seminary who correctly harped on the subject: "Edit and Revise;" Emil R. Sterbenc, a retired business executive with mastery of the English language; Garland E. McBride, a retired newspaper executive who cautioned against verbosity; and cousin, C. Milton Kelsey, a retired business executive, who admonished me for straying too far afield. All these wonderful people, and many others, materially assisted with the overall composition while ensuring that the text would be understandable to lay readers.

Appreciation is extended to Dr. Joseph H. Davis, the Chief Medical Examiner for Dade County, Florida. I am grateful to Dr. Davis for the authorization to use his name, and the exerpts from his expert opinions, that have been reprinted in the book.

I am indebted to Ronald K. Martin, Managing Editor of the Florida Times Union, Jacksonville's daily newspaper, who granted authorization to use the two news stories from their papers that are reprinted in Chapter Eighteen.

Special recognition, and my sincerest appreciation, are extended to those three very special individuals who contributed so much to this endeavor. Although not specifically identified, you know who you are, and I shall forever hold you in the utmost regard, and everlastingly remember you with the highest esteem.

The cooperation rendered by Bette R. Jones was indispensable. Besides assisting with the design of the cover and typesetting the manuscript, the adoption of her editorial suggestions significantly improved this book.

I would be remiss if I failed to recognize Barbara M. Clark, the grammarian and proofreader of the completed text. Looking over her notes caused me to regret very much my lack of attention during English classes. In concert with other corrections, Ms. Clark undangled the participles, reunited the infinitives, and created order out of chaos when she rectified my helter-skelter punctuation.

TABLE OF CONTENTS

APPENDIX

FOREWORD

The conduct of complete and thorough investigations of transportation accidents by an organization having no vested interest whatsoever in the ultimate outcome was one of the guiding principles behind the creation of the National Transportation Safety Board. Following its in-depth investigations, the Safety Board's probable cause determinations correctly cite the factors involved somewhere in the 98th to 99th percentile. However, on many occasions during my investigations I heard the statement: "All you do is blame it on the pilot."

It is true the Safety Board ascribes about 80 percent of our aviation accidents to pilot factors. And rightly so, because, regardless of the field of endeavor, people cause most accidents. From an investigative standpoint, covering a pilot-caused accident is not all that exciting. Similar accidents occur repeatedly. Essentially, the only things that change are the registration numbers of the airplanes, the names of the people involved, and the locations of the accidents.

We have to realize it is impossible to manufacture aircraft, train pilots, operate our air traffic control system, write regulations, etc., so as to prevent all mistakes by fallible mankind. Conceivably, that could require another Sigmund Freud who would be capable of climbing inside our heads and tightening all the loose screws.

Basically, it is our temperament that accounts for our actions. Temperament relates to disposition, frame of mind, nature, personality, etc., and everyone knows there are vast differences in those characteristics amongst Homo sapiens. In "The Turning-Point in My Life," Mark Twain compares Adam and Eve to Martin Luther and Joan of Arc, to illustrate different temperaments and show how they affect behavior. Citing Martin Luther and Joan of Arc's temperaments as being made of "asbestos," he wrote: "By neither sugary persuasions nor by hell fire could Satan have beguiled them to eat the apple."

After due study, Dr. Freud would be able to determine which of the diverse temperamental groups we are associated with. That would explain why some of us can serenely glide along the surface of life creating hardly a ripple; while others, with different temperaments, are so truculent, accident-prone, and full of hell fire. However, it is doubtful whether Dr. Freud could tell us how to escape our intrinsic temperaments to a degree sufficient to move us from one category to another.

So, insofar as the subject matter of this manual is concerned, it is considered Mark Twain's contentions are more appropriate. We are what we are, and that explains why the same types of accidents continue to be repeated time and again.

INTRODUCTION

In the beginning they finally figured out how to build the iron bird, and a marvelous achievement it was. But then a pilot cracked up his flying contraption, and that's the way it went. They built them and smacked them up until a smart somebody decided there was little sense in having so many mishaps without figuring out what was causing them. So, safety bodies came into being, that, insofar as the United States is concerned, have evolved into the present day concept of the National Transportation Safety Board.

Aircraft accidents producing many tombstones always receive wide media coverage. As the evidence is being gathered, the Safety Board keeps the public aware of the factors that might have been involved. Then about a year after the mishap, the Safety Board publishes an in-depth report containing the facts, circumstances, and conditions of the accident. These so-called "Blue Cover" reports are loaded with technical matter beyond the grasp of the lay reader. Therefore, the public normally accepts the Safety Board's probable cause determinations while relegating the remainder of the lengthy reports to the aviation community, and those ever-present liability lawyers.

For the general aviation, business aircraft, and less serious air carrier accidents, the Safety Board spends innumerable hours feeding statistical data into its computers. Then about twelve times a year, the computer coughs out a book of accident releases in a brief format style that includes the Safety Board's probable cause determinations. The releases are loaded with "computerese" jargon, and the circumstances of an accident just don't **"FLOW,"** as the reader is forced to hop all over the page to determine the **HOW, WHAT,** and **WHY.** All of which makes for pretty dull reading except for individuals or organizations having a personal, or vested, interest in a particular mishap.

This book attempts to correct those deficiencies. Although the reporting of the facts, circumstances, and conditions for so many different types of aviation accidents in one volume will benefit the aviation community, it has been adapted for the general public through use of understandable, nontechnical, language. Besides covering specific occurrences, this digest provides a brief history of the Safety Board, plus insight into its mode of operation and the investigative process.

Wishing all a Happy Landing.

Tom Watson

"It is unwise, my dear Watson, to speculate in advance of the facts," Holmes admonished. "Invariably, it biases the judgment."

Sir Arthur Conan Doyle

MISLEADING INFORMATION

It was a snowy afternoon with subfreezing temperatures at an airport serving the nation's capital. The airport had been closed from 1:30 p.m. to 2:50 p.m. for snow removal. While most major airports have multiple runways permitting snow to be removed from one while operations are continued on another, this closing was necessary because the airport had only one instrument runway.

While the airport was closed, incoming traffic stacked up overhead. Because of the limited number of terminal gates, all manner of congestion was created on the airport's restricted ramp space when departing planes had to be pushed back early in order to provide gates for arriving traffic to deplane passengers. Adding to the confusion was the requirement that departing flights had to be deiced without encountering extensive delays before taking off that would permit the ice and snow to reaccumulate on the airplanes. Ideally, the traffic flow at the airport is spaced to permit a departure between each arrival.

Ten minutes after the airport was reopened, but about 50 minutes after the airplane had been deiced, the flightcrew received takeoff clearance. The copilot was flying the airplane, and the captain was handling the throttles, monitoring the engine gauges, etc. While the airplane was accelerating down the runway at a slower than normal pace, and during the initial climb after lift-off, the transcription of the cockpit voice recorder tape showed the following conversation between the flightcrew:

> **Copilot - "God, look at that thing."**
>
> **Copilot - "That don't seem right does it?"**
>
> **Copilot - "Ah, that's not right."**
>
> **Captain - "Yes it is, there's eighty." (knots airspeed)**
>
> **Copilot - "Naw, I don't think that's right."**
>
> **Copilot - "Ah, maybe it is."**
>
> **Copilot - "I don't know." Followed by the sound of the stickshaker (a device that activates to warn the crew of an impending stall) which continued until the airplane crashed.**
>
> **Copilot - "Just barely climb."**
>
> **Copilot - "Larry, we're going down, Larry."**
>
> **Captain - "I know it."**
>
> **Thereafter, the sounds of impact were recorded as the airplane crashed.**

No, this accident didn't happen in some developing third world nation with limited resources. As many readers have already guessed, this was Air Florida's Flight 90, a Boeing 737 airplane, that was departing from Washington's National Airport en route to Fort Lauderdale, Florida, with an intermediate stop in Tampa, Florida.

The airplane struck the heavily congested northbound span of the 14th Street Bridge, which connects the District of Columbia with Arlington County, Virginia, and plunged into the ice-covered Potomac River. When the airplane impacted the bridge, it struck six occupied automobiles and a boom truck, before tearing away a 41-foot section of the bridge wall and 97 feet of the bridge railing. Seventy airplane passengers and four crewmembers were killed. Four passengers and one crewmember were injured. Four persons in vehicles on the bridge were killed, and four others were injured.

The aft fuselage split open during impact just forward of the vertical fin. Through this opening, six occupants escaped, including a flight attendant who inflated the only available life vest and gave it to a more seriously injured passenger. One of these six eventually drowned.

A U. S Park Police helicopter arrived at the scene about 20 minutes after the crash. Although not specifically equipped to conduct water rescue operations, the crew predominated in the rescue effort. The pilot hovered near the survivors while his crewman dropped makeshift survival aids, consisting of ropes with loops and life rings, to the survivors in the water. With the survivors hanging onto the ropes and rings, the pilot dragged them to the shore. To accomplish one rescue, the crewman stood on the helicopter's landing skid and pulled a survivor from the water. Bystanders on the shore assisted with the rescue efforts and one, exhibiting total disregard for his own safety, jumped into the frigid water and swam to the aid of one of the survivors.

Circumstances associated with the survival of the four passengers and one flight attendant produced enough real life heroes and heroines to keep Hollywood busy for a goodly spell. For the possible reason that there were no realistic ways to inject drugs, sex, violence, an automobile chase, or cops and robbers, into the plots, producers have failed to capitalize on the opportunity. However, the heroic and selfless efforts of those who contributed to the rescue of the survivors did not go unnoticed by the Safety Board, the news media, or local authorities.

Belittling the performance of airport authorities is not intended. It is impossible to construct more terminals, provide more ramp space, and add additional runways when adequate real estate for such projects is not available. Because of its close proximity to the nation's capital, the National Airport is very popular with travellers. Passengers arriving at Washington's National airport are spared the rather lengthy automotive trips from the "safer" Dulles International Airport in Virginia, or the Baltimore-Washington International Airport in Maryland. Notwithstanding the limited facilities that provide little margin for error, the National Airport has a very commendable safety record. For this, credit should be extended to the air traffic controllers for devising a system to deal expeditiously with the heavy traffic flow, and to the flightcrews who continue to exhibit such fine airmanship under unfavorable conditions.

When asked about the cause of this accident, most people have expressed the opinion that it was the ice and snow on the airplane's wings. However, the Safety Board's final report released seven months after the accident showed the predominating factor in the accident was the decreased engine thrust used for takeoff. Although other issues were involved, the accident would not have happened if the crew had used the correct engine thrust settings on the takeoff.

It is the Safety Board's policy to release factual information to the public as it's gathered. While this policy probably has some merit, it oftentimes demonstrates that there is much truth in the great detective's statement at the beginning of this chapter. After such a catastrophic occurrence, the news media are literally begging for any information they can obtain. In this instance, the initial factual information basically concerned the ice and snow that had accumulated on the airplane. Additionally, there was much discussion about the deicing of the airplane (removing ice and snow clinging to the outer surface) and the ensuing delays before takeoff that might have contributed to a reaccumulation of the ice and snow on the wings. While buckets of this information were being fed to the public by every available media means, other, more time-consuming aspects of the investigation were ongoing.

In keeping with the saying, "there's nothing older than yesterday's news," the public's interest in this disastrous accident tapered off day by day. By the time the important information concerning the use of reduced engine thrust had been gathered, confirmed, and released to the media, most folks took little notice, or otherwise entirely missed its significance. However, they still have almost instant recall of all the talk about the ice and snow on the airplane reported at the outset of the investigation.

This was an extremely interesting accident and many issues resulting in improved crewmember performance were uncovered. However, plowing through the Safety Board's 164-page report of the facts, circumstances, and conditions can become a very laborious task. Likewise, the wording in the Safety Board's probable cause determination may not be fully understood by the general populace:

> **"The National Transportation Safety Board determines that the probable**

cause of this accident was the flightcrew's failure to use engine anti-ice during ground operation and takeoff, their decision to takeoff with snow/ice on the airfoil surfaces of the aircraft, and the captain's failure to reject the takeoff during the early stages when his attention was called to anomalous engine instrument readings. Contributing to the accident were the prolonged ground delay between deicing and the receipt of ATC takeoff clearance during which the airplane was exposed to continual precipitation, the known inherent pitchup characteristics of the B-737 aircraft when the leading edge is contaminated with even small amounts of snow or ice, and the limited experience of the flightcrew in jet transport winter operations."

What a mouthful! It might even be said they left no turn unstoned. On some occasions, it is difficult to fathom why the Board Members are reluctant to just go ahead and call a spade a spade, or a bad name, or whatever, and be done with it. However, although it takes some digging, and possibly some basic knowledge of the issues involved, the pertinent facts, circumstances, and conditions relating to the accident were included in the Safety Board's report.

The crux of the matter is the flight was lost because the flightcrew failed to utilize the engine anti-ice system, and this resulted in faulty engine thrust readings. Regardless of the ice and snow accumulation on the airplane, the delays before takeoff, any pitchup characteristics of the airplane, or the flightcrew's experience level in winter operations, they would have experienced no difficulty whatsoever with the takeoff and climbout if they had used the predetermined engine thrust settings. Likewise, even after liftoff, the accident could have been prevented if the flightcrew had just increased the thrust levers sufficiently to obtain the desired thrust output.

The airplane was equipped with two turbo-fan engines having takeoff thrust ratings of 14,500 pounds each. The engine instrument cluster in the cockpit contained five separate instruments for each engine. The engine pressure ratio gauge readings are primarily used by the flightcrew to obtain the desired thrust output from the engines. The ratios are obtained by dividing the compressor inlet pressure at the nose into the turbine discharge pressure at the rear of the engine. The ratios are determined electronically and displayed continuously on the engine pressure ratio dials on the cockpit instrument panel.

The flightcrew intended to use 2.04 pressure ratio readings on both engines for takeoff. When icing conditions exist, use of the engine anti-ice systems prevents the inlet pressure probes in the engine noses from becoming blocked by ice. In those cases where the probes do become blocked by ice, as in this instance, lower than existing compressor inlet pressures are used in the pressure ratio computations. Accordingly, lower than desired turbine discharge pressures were required to obtain what were actually faulty 2.04 pressure ratio readings. Safety Board investigators determined the flightcrew's use of the faulty 2.04 pressure ratio indications resulted in the actual use of about 1.70 pressure ratios for the takeoff. This caused a takeoff thrust deficiency of about 3,750 pounds per engine, which is slightly more than 25 percent of the engine's takeoff thrust ratings. Knowledgeable aviation personnel will tell you that's a whole bunch of thrust to go unused on any takeoff.

The cockpit instrument panel also has two revolutions per minute (rpm) gauges for internal engine components, exhaust gas temperature gauges, and fuel flow gauges that show the fuel consumption for each engine in pounds of fuel per hour. Obviously, the higher the rotational velocities, exhaust gas temperatures, and fuel flows, the greater the thrust output from the engines.

The copilot's remarks during the takeoff were related to the engine instrument readings. Although the pressure ratio gauges showed both engines were developing the desired thrust, there were deficiencies in the readings of other engine gauges. The engine pressure ratio gauges are the primary thrust-setting instruments, but they should never be relied upon singularly as GO-NO-GO indications by the flightcrew. Instead, a crosscheck of the other engine instruments is required to show that they are also indicating the proper parameters. The crosschecks are necessary to verify that the pressure ratio readings are valid. This is called instrument scan, and is identical to that used in instrument flying conditions wherein the artifical horizon or flight director is the primary flight attitude-indicating instrument. However, a constant cross check of the airspeed indicator, altimeter, rate of climb indicator, direction indicator, etc., is always required.

The Safety Board arrived at 37 separate findings in its analysis of the accident. The selected findings below zero in on the pertinent factors involved in the accident:

> **1. The flightcrew did not use engine anti-ice during ground operation or takeoff.**

2. The engine inlet pressure probes on both engines became blocked with ice before initiation of takeoff.

3. The flightcrew set takeoff thrust by reference to the engine pressure ratio gauges to a target indication of 2.04, but the pressure ratio readings were erroneous because of the ice-blocked probes. The copilot, who was flying the airplane, was aware of the disparity between the pressure ratio indications and the readings on the other engine instruments. While not addressing the subject directly, he expressed concerns that something was "not right" on four occasions during the takeoff roll; however, the Captain took no action to reject the takeoff.

5. The Safety Board concluded that even with the low thrust during the takeoff roll and the aerodynamic penalty of the snow and ice contamination, the accident was not inevitable as the aircraft lifted off. However, both immediate recognition of the situation and positive effective actions by the flightcrew to counter the noseup pitching moment, and to add thrust, were required.

The news release accompanying the distribution of the Safety Board's report of the accident contained the following: "The Board said it believes that the flightcrew hesitated in adding thrust because of their concern for engine limitations, but added that the crew should have known all other engine indicators were well below limit values."

I investigated many other occurrences that terminated catastrophically because of the basic tendency on the part of some pilots to avoid damaging their airplanes at all costs. Based on an analysis of the factual gatherings in those instances, it was evident some pilots are prone to do everything possible to prevent "bending the bird," or otherwise damaging their flying machines, even at the expense of safety for everyone on board. While admitting that I do not possess the requisite background to fully analyze and explain the psychological factors involved, the pilots' attempts to save their airplanes were considered inherent to their basic behavior.

On the other hand, there are pilots whose attitudes towards their airplanes are, "I'll take care of this airplane as long as it takes care of me." A pilot endowed with such a posture would have jammed those thrust levers all the way forward before allowing his airplane to descend into the 14th Street Bridge.

Although possibly exposing some cowardice on my part, it was demonstrated in my flying days that I am imbued with the latter inclinations. During the Korean war, while assigned to a squadron flying jet fighters equipped with engines utilizing centrifigal compressors, we went through a period where we experienced an inordinate number of engine failures due to compressor disintegration. Everyone should realize that the higher revolutions required to obtain maximum thrust from the engines created the greatest centrifigual stresses on the compressors. Maximum power was required for takeoffs with our heavy fuel and ordnance loads. However, after becoming airborne, we were advised not to exceed 95 percent in order to reduce the centrifugal loads on the compressors.

This posed no problems. However, on the next several pullups after bombing or strafing runs, I found myself using 100 percent again. A conscious effort on my part to avoid the practice was ineffective, especially when the air was filled with flak and visible tracers. Thereafter, I readily admitted making more than one takeoff on each mission, one from the airfield and the others on pullups from the target area. I reckoned there was more likelihood the compressor would hold together than there was that the enemy gunners would miss their target.

> ***The Lord God said to Adam; "Have you eaten from the tree that I commanded you not to eat from?" Adam replied; "That woman you put here with me, she gave me some fruit from the tree, and I ate it." Then the Lord God said to Eve; "What is this you have done?" And Eve replied; "The serpent deceived me, and I did eat."***

- -

THE INFALLIBLE BREED

Most readers are familiar with the above scripture taken from the third chapter of Genesis. It is used herein to illustrate that the inclination to blame our faults, misdeeds, shortcomings, mistakes, or other sins of commission or omission on something or somebody else, could well be an inherent manifestation of human behavior that has been with us since the very beginning of human existence on this planet. And, as we shall learn, the tendency to point our fingers at something or somebody else can become even stronger when it involves an individual or group that has a vested interest in the outcome.

A McDonnell-Douglas DC-9 jet transport airplane literally broke its back when the upper fuselage split open during a hard landing. The airplane came to a stop on the runway, resting on the landing gear, but with the bottom of the aft fuselage dragging the runway pavement. The accident occurred at night during moderate rain showers. The ceiling and visibility were well above the prescribed minimums for the flightcrew's instrument landing system approach to the runway. The Safety Board assembled what is commonly referred to as a partial team effort, with the Investigator-in-Charge being assigned from its Miami Regional Office and necessary technical assistants, under the titles of Group Chairmen, from its Washington, DC, headquarters.

My assignment only related to the writing of a report for approval and adoption by the members of the Safety Board. I was not involved in the conduct of the on-scene or subsequent phases of the investigation. However, when the report writer completes his assignment, he knows as much about an occurrence as anyone else. It is necessary to study the various group chairmen reports, crewmember statements, air traffic control data, company operating procedures, the cockpit voice recorder and flight data recorder readouts, etc. All manner of evidence is collected during the course of an investigation. Before writing the report, it is necessary to determine the evidence pertinent to the factors involved that are to be highlighted in the accident report.

The captain was flying the airplane. Information obtained from readout of the flight data recorder showed he made a smooth approach with a fairly constant airspeed and rate of descent until shortly before touchdown. At that point, the approach became destabilized, with a resultant increase in airspeed and rate of descent. Readout of the cockpit voice recorder tape showed that the copilot warned the captain of the excessive rate of descent just prior to the touchdown. The statement by the cockpit jump seat occupant, another company McDonnell-Douglas DC-9 copilot, showed he also noted an excessive sink rate just prior to touchdown. The captain gave a comprehensive statement, but said he never thought his rate of descent was excessive.

The airplane was equipped with a ground proximity warning system (GPWS) that is designed to provide the flightcrew with aural and visual warnings of five hazardous conditions. Mode 1 provides a warning of excessive sink rate. During playback of the cockpit voice recorder, "Pull Up" warnings triggered by the GPWS were heard about three seconds before ground contact. Additionally, tripped components on the GPWS showed the airplane contacted the runway while in a rate of descent that exceeded the manufacturer's design limitations.

A comprehensive review of all the data contained in the accident file showed this was basically a pilot factor occurrence. It was recognized that a lower than stipulated tension on the captain's windshield wiper arm might possibly have created some visionary illusions in the moderate rain, but the captain said he didn't consider that to be involved. Believing we had a pretty firm grasp of the accident and the factors involved, we were rather frantically striving to meet our deadline for submission of the initial draft to our Washington headquarters. Two days before the deadline, the Investigator-in-Charge came into my office and handed me information he had received in the morning mail from the Air Line Pilots Association (ALPA.)

The information related to medical data concerning the captain. It showed that, subsequent to the accident, he had been diagnosed as having Parkinson's disease. Attached thereto was a report by the Emory Medical Clinic in Atlanta, Georgia, showing the captain had noticed a peculiar sensation in his left shoulder and arm for the past two years. The data also included the results of a special aeromedical evaluation at the Mayo Clinic in Rochester, Minnesota, showing the captain had noticed some increasing fatigue and weakness in his left arm for several years. A cursory review of the medical information showed we would need copies of the captain's medical records on file with the FAA in Oklahoma City, Oklahoma, before we could complete our report.

Most folks probably have a favorable opinion of the Air Line Pilots Association. Its representatives have little difficulty being interviewed on TV when some grave or pressing aviation matter is up for discussion or review. The spokesmen are impressive looking, neatly dressed, knowledgeable about the subject, and routinely leave the viewer with the opinion that ALPA's primary interest is aviation safety. However, there are many Safety Board investigators, FAA inspectors, air carrier employees, and others concerned with aviation matters who, like the author, realize ALPA has other high ranking considerations.

We need to understand that the Air Line Pilots Association is a union organization. Obtaining and holding a favorable public opinion for the association ranks high on its priority list. Over the years, that goal has been achieved quite effectively with an information campaign that has led the public to believe its principal interest is aviation safety. Another important objective concerns the protection of the membership. During the conduct of aircraft accident investigations where pilot factors are indicated, its representatives sometimes attempt to accomplish this objective by finding or suggesting mitigating circumstances that could have influenced crewmember performance, or by bringing to light other circumstances or events that might steer the thrust of an inquiry away from pilot factors.

In 1988, reports began to surface of instances where flightcrews involved in accidents were sequestered for inordinate periods of time before being available for interview by those conducting the investigation. When discussing these events with other investigative personnel, it was concluded the practice was designed to sanitize statements by the flightcrews insofar as pilot factors may have been involved in the cause of an accident.

Over the past decades, airline safety has greatly improved, and the Air Line Pilots Association materially contributed to those accomplishments. However, from my viewpoint, ALPA's inputs to this ongoing program were made with safety ranking no higher than level par on the association's priority list alongside its public image and the protection of the union's membership.

Let's put this in perspective by reviewing some past events. The pilots of one of our major domestic air carriers go on strike. There have been occasions where the pilots of other air carriers employed slowdown tactics insofar as their particular company's operating procedures, and their involvement in the air traffic control system, were concerned. Anything that greatly inconveniences the general public is quite newsworthy. Air Line Pilots Association spokesmen routinely espoused safety when queried about these actions. But their answers were not necessarily true. Indeed, there is reason to believe it was a union tactic designed specifically to disrupt the orderly flow of air traffic, with the expectation that it would influence the proceedings at the bargaining table. If ALPA's primary interest was safety, it would insist that the same slowdown tactics be incorporated into the nonstrike, day-to-day operational procedures.

Most Safety Board investigators are aware, and many lay readers might conclude from the above, that participation by the Air Line Pilots Association in aircraft accident investigations where pilot factors are involved is not always beneficial. Oftentimes, it seems that ALPA's representatives are out there throwing banana peelings in your path rather than objectively assisting with the orderly collection of the pertinent evidence concerning the facts, circumstances, and conditions of an accident. However, although unintentional on its part, participation by the Air Line Pilots Association is advantageous because it ensures that the Safety Board's investigators do a far better job than should actually be required. By the time all the union's tactics and inputs designed to steer the course of an investigation away from crewmember performance are investigated and evaluated, a most thorough and complete inquiry has been accomplished.

Back now to the pilot with Parkinson's disease who broke the back of his McDonnell Douglas DC-9 airplane while landing. It was obvious to us that the Air Line Pilots Association was trying to come up with some mitigating circumstances to account for the captain's performance. While reviewing with the Investigator-in-Charge the medical records from the Emory and Mayo Clinics, we realized the Air Line Pilots Association's tactic would almost certainly backfire, and upon final analysis, it might even appear that ALPA had shot itself in the foot. This assessment was based on our knowledge of the FAA's medical standards. In the event the captain was experiencing any symptoms that would make him unable to meet the physical requirements of his current medical certificate, it was his duty, responsibility, and obligation to take himself off the flight schedule until he had consulted with an authorized FAA-designated medical examiner in compliance with existing regulations.

The copies of the captain's medical records we received from the FAA showed he made five applications for medical certificates during the 26-month period prior to the accident. His applications listed no changes in his medical condition and he reported no illnesses. Parkinson's disease begins insidiously, with any of its three cardinal manifestations characterized by tremor, muscular rigidity, and a loss of postural reflexes, either alone or in combination. Tremor usually in one or sometimes both hands, involving the fingers in a pill-rolling motion, is the most common initial symptom. This symptom is often followed by stiffness in the limbs, general slowing of movements, and inability to carry out normal and routine daily functions with ease.

The Safety Board concluded that the early symptoms being experienced by the captain at the time of the accident had not progressed sufficiently to adversely affect his ability to fly the airplane. Nonetheless, the Safety Board expressed concern that the captain withheld information regarding these early symptoms on his recent applications for FAA airman medical certificates.

HAULED INTO COURT

After taking off from the Greenville Downtown Airport in Greenville, South Carolina, the pilot of a light, twin-engine, commuter, passenger/cargo flight was advised by the tower controller his airplane was streaming smoke from the left engine. The pilot replied: "Okay, I'm coming back in." He commenced a left turn, but remarked to the passenger occupying the right cockpit seat that they wouldn't be able to make it. Thereafter, the airplane collided with power lines adjacent to a busy thoroughfare and crashed into the parking lot of a Steak and Ale restaurant. Post impact ground fire occurred and the wreckage was fire gutted.

The pilot and four passengers were killed and the other two passengers sustained serious injuries. The members of the team assembled for the investigation possessed impressive credentials. It included two FAA inspectors from the Columbia, South Carolina General Aviation District Office; two inspectors from the FAA's Southern Regional Headquarters in Atlanta, Georgia; two powerplant specialists from the Safety Board's Bureau of Technology in Washington, D.C.; and representatives from: Piper Aircraft Corporation, the airplane manufacturer; Lycoming Engine Division, the engine manufacturer; the Bendix Corporation, whose components were installed on the engines; and AiResearch Company, who manufactured the engine turbochargers.

While I was having breakfast with the FAA coordinator at our motel in Greenville, an individual introduced himself, handed me his business card, and said he would be representing the operator during the conduct of the investigation. A glance at his card showed his company was a fully owned subsidiary of an insurance group. I said, "Fred, you're trying to wear too many hats; you cannot participate or represent any organization as a party to this investigation because of your affiliations with the insurance group."

He replied something to the effect that if we didn't let him participate, the operator planned to shut down the investigation. I responded, "That's not my concern, because the Safety Board's regulations preclude me from designating you a party to the investigation." However, that's just what was said outwardly. Inwardly, I was longing for the skirmish. All of us probably enjoy doing battle on occasion. "Doing battle" as used herein doesn't necessarily have reference to brawn and muscular agility. It is readily admitted I'm rather ill-equipped in both those departments. However, being a member of that group who cherish the opportunity to be taken to task on some issue they fully understand, I was looking forward to being dragged into a battle of wits, if that's what it might be called. My response was prompted by knowledge of the Safety Board's regulation that states the following: "No party to the field investigation shall be represented by any person who also represents claimants or insurers."

The accident happened on Friday, and by Sunday we had completed our activities at the crash site and moved the wreckage to a hangar with plans to commence the detailed examinations of the engines on Monday. While resting in my motel room that evening, I responded to a rap on the door and was handed a court order suspending further investigative activities. This was a very newsworthy occurrence in the local area. My somewhat facetious response to a reporter who asked what we intended to do didn't exactly sparkle with intelligence. He was told I had long ago established a firm policy of always complying, to the very letter, with all court orders. This was another lesson learned: that you have to watch what you say in such instances if you don't want to hear it again on the evening news or read about it in the morning papers. The basis for the court order was my refusal to name as a party to the investigation the individual affiliated with the operator's insurance company.

Large telephone bills were run up that evening while I was briefing my supervisor in Miami, Florida, followed by a detailed discussion of the events and circumstances with department heads in our Washington headquarters. Preservation of the Safety Board's authority to name or exclude parties to its investigations was almost sacred. The loss of that prerogative would mean loss of control for the overall conduct of the investigative proceedings in many instances. Congress, in one of its rare displays of wisdom, had granted automatic authority only to the FAA to participate in Safety Board investigations. Regardless of the level of expertise other personnel or organizations might possess, their participation is based on the Safety Board's need for such expertise. It all boils down to a judgment call by the Investigator-in-Charge, who has the authority to name, or exclude, various parties who might desire to participate.

The response by our Washington staff was swift and positive. By midmorning on Monday, I had secured the investigation and was headed for Columbia, South Carolina, to meet with an assistant U. S. attorney assigned to the case. A Safety Board attorney was also en route from our General Counsel's Office in Washington, D.C. The assistant U. S. attorney was familiar with the local court scene and had knowledge of the customs and procedures of the presiding federal judge who had issued the restraining order. He had me prepare an affidavit relating to the conduct of the investigation until halted by the court order and the names of those designated as parties to the investigation. Attached to the affidavit were copies of the Safety Board's applicable regulations and procedures. Basically, our side of the issue would be based on my testimony concerning the accident in question plus a defense by our lawyers for preservation of the Safety Board's authority to exclude persons or organizations with certain affiliations as stipulated in the regulations.

In a surprising example of hasty action by our court system, the hearing was scheduled for Thursday. The size of the entourage and the voluminous files the plaintiff's side brought into the courtroom were puzzling. Apparently, they were eager to air their side of the issue and had made preparations for lengthy proceedings. We all stood in the courteous display of respect as the Honorable Judge Robert Chapman entered the courtroom. He wasted little time and after everyone was seated, announced that this was the case of the plaintiff versus the National Transportation Safety Board. He asked the plaintiff's side if they were prepared to proceed and they answered in the affirmative. He received a like reply to the question from our side. The remainder of the courtroom proceedings are forever etched in my memory.

The Honorable Judge stated that before proceeding further, he had several questions. He asked our side about the status of another individual with questionable affiliations who had been observed at the crash scene. Our attorney replied that the individual would not be permitted to participate in the investigative proceedings. The judge then asked the plaintiff's side who employed the rejected party. They gave the name of the company and the judge asked if that was an insurance company. They responded in the negative and the judge asked who owned the company. They gave the initials of the company and the judge asked if that was an insurance company. They gave an affirmative reply.

The judge's response was quite dramatic and final. He ruled the rejected party could not participate, saying the issue was "as plain as the nose on my face." He found the Safety Board's regulation prohibiting the individual's participation to be valid. He lifted the restraining order which he said he wouldn't have signed in the first place if he had known the party worked for an insurance company. And with that, Judge Chapman rose, and walked out of the courtroom.

The assistant U. S. Attorney for our side turned to me and said: "Tom, if you want to observe something interesting, go out in the hallway and watch while the bevy of attorneys from the other side retry the case over and over again." He remarked that it happened quite frequently and would probably occur in this instance because of the extremely brief session. I took his advice and was amused by their antics as they pointed out to each other the errors that had been committed in the courtroom and the overwhelming weight of their evidence relating to the matter, if only they had been allowed to present it. While I was leaning against a wall in the corridor, giving a poor performance of hiding my euphoric feeling and sense of satisfaction brought on by the judge's ruling, one of the attorneys from the other side came over and rather caustically asked what I'd think if they appointed Professor (name deleted) to represent the operator. I reminded him that we had just established what is called a legal precedent in his line of endeavor, and the Safety Board would be waving about and leaning heavily on the judge's decision for a long time to come. Therefore, for the duration of the current investigation, we didn't care if they appointed a three-headed monkey as long as the designated party didn't also represent claimants or insurers. In reality, however, we weren't so naive as to believe anyone they named at this stage of the proceedings wouldn't actually be representing claimants or insurers in some capacity.

It is important to understand that designation of an individual to represent a particular company or organization as a party to an investigation is a two-way street. While such individuals have access to the evidence as it is collected, they are obligated to provide the Investigator-in-Charge with the knowledge and expertise their organization possesses that will benefit the proceedings. Very much of a spirit of cooperation and an aura of camaraderie prevails while the methodical and time-consuming collection of evidence is conducted under the overall direction of the Investigator-in-Charge. Throughout the proceedings, the various parties engage in free and open discussion of any evidence collected that might be pertinent to the accident.

The individual named by the attorney to represent the operator after the court session exhibited all the classic qualities and characteristics of a "hired gun," a title customarily given to parties not interested in any methodical collection of evidence. Instead, they are looking for anything out of the ordinary to take into the courtroom for the apparent purpose of muddying the issue without regard as to whether it may have had any bearing whatsoever on the accident. They are readily recognizable because they stand out like a turkey gobbler in a flock of chickens. No warnings or cues were required to halt the open discussions among the various parties that were being conducted before the professor came on the scene. Everyone automatically ceased the in-depth discussions of the issues, to avoid providing data for the "hired gun" to latch onto. Oftentimes, a "hired gun" will want the investigation to go off on some tangent or explore some avenues that couldn't possibly have contributed to the accident. Such requests never posed any problems. My routine response was: "Not as a part of or during the Safety Board's investigation." If they wanted to conduct unnecessary tests on components or accessories, they were advised: "You'll have to wait until the wreckage is released from the Safety Board's custody before you can make arrangements to further examine, test, or disassemble, those components."

Under normal circumstances, the actions of the professor would have provided justification for his dismissal from the ongoing proceedings because he was not being objective. Safety Board regulations state: "Participants in the field investigation shall be responsive to the direction of the appropriate Board representative and may be relieved from participation if they do not comply with their assigned duties or if they conduct themselves in a manner prejudicial to the investigation." In this instance, however, with the favorable court decision in hand, we tolerated his participation while closely monitoring his activities.

There was renewed local interest in the accident after the court session. Our Washington headquarters was requested to provide sufficient funds to put on what I refer to as a "Dog and Pony Show" for the news media. They granted the request and we conducted the engine teardown examinations at the Stevens Beechcraft facilities at the Greenville-Spartanburg Airport. Stevens Beechcraft was selected because they had a very clean, organized, and impressive maintenance hangar.

The news media were granted whatever access they desired to the engine examinations and, with all the cumulative expertise, we put on a fairly respectable show, albeit we didn't find the Easter Egg. "Easter Egg" is a term used by investigators to define the precise thing or event that caused an accident.

Some discrepancies were noted during review of the airplane records, and discussions with those familiar with the airplane and its maintenance history revealed evidence of inadequate or improper maintenance and inspection. Most of the engine accessories were destroyed in the postcrash fire. Five cylinders on the left engine showed evidence of an excessively rich mixture, and the remaining cylinder showed evidence of an excessively lean mixture. While this finding was considered significant, complete disassembly inspection of the engine and functional tests of airplane and engine components that were not destroyed failed to provide conclusive evidence as to the factors related to the pilot's inability to remain airborne.

It was always my opinion that nothing could beat a fully functional test when attempting to determine the precrash operational capability of various components. The right engine fuel injector component was destroyed in the postcrash fire, but the one on the left engine appeared to be intact. Efforts are always exerted to test components after the least possible degree of disturbance. However, because of the postcrash fire in this instance, the component was removed and hand carried to the engine manufacturer's facility in Pennsylvania.

After installing the fuel injector on an engine in a test cell, the technician was requested to conduct the tests he routinely performed. At the conclusion of the tests, he stated all readings were within the prescribed parameters. I requested that he shut down the engine and remove the component. I refused the professor's request for permission to make some adjustments on the component to determine how they would affect its operation. I further explained that any additional tests would not be a part of the Safety Board's investigation. The professor was informed we had already obtained permission from the owner to release it to his custody, and after handing the fuel injector over to him, we went our separate ways.

Several weeks later he called to tell me what they had discovered while conducting further tests on the component. He was advised we weren't interested. His request that we delay submission of my factual report for the accident was refused. In response to his query as to how he might get his findings into the record, he was informed the only avenue would be a written petition to the Safety Board in Washington. Thereafter, I had no further contact with the individual.

To this day, I still defend my actions. Once the component passed the operational tests, nothing further that might be pertinent to the cause of the accident could be found. Anything that was found thereafter would only provide material for hypothetical conjecture, and if the Safety Board allowed that to go on indefinitely, we would never be able to complete some investigations.

The accident was a classic example of those occurrences where it is impossible to determine the cause precisely. The Safety Board's probable cause determination showed powerplant failure for undetermined reasons and cited the inadequate maintenance and inspection and the high terrain obstructions as factors in the accident. There is no question that significant evidence was lost during the crash sequence and postcrash fire. All the experts assembled for the investigation couldn't come up with the reasons why the engines didn't generate sufficient power for the flight to remain airborne, and the Safety Board never resorts to guesswork in its probable cause determinations.

Some occurrences result in near pen-pal relationships between the Investigator-in-Charge and others having an interest in an investigation. The term "pen-pal" is used rather loosely because, in this instance, it makes reference to an individual who literally became "a thorn in our side." We're referring to persons who absolutely refuse to accept "NO" for an answer to their queries and requests. Such was the relationship between myself, and ultimately the Safety Board in Washington, DC, and a Greenville, South Carolina, attorney representing liability claimants after this accident. I am a member of the group that gives an emphatic "NO" when warranted by the circumstances. However, our folks in Washington exhibited more savoir-faire, and were prone to mollycoddle petitioning individuals.

The attorney's letters started on a harmonious note, but became more sarcastic when his communiques failed to achieve the desired actions. He ultimately had some very unsavory things to say about the overall conduct of the investigation and chastised me pretty severely for "shooting off my mouth to the press" when some of the factual evidence obtained during the investigation was released. However, it should be mentioned this practice is encouraged by the Safety Board.

Basically, the thrust of his request was that we reinterview a surviving passenger, claiming he was "in shock due to having gone through such a traumatic ordeal" when we initially conducted our interview. I had gained some lay knowledge in this arena from the conduct of many prior interviews with passengers and aircrewmen who had been injured in accidents. In this instance, both the accompanying FAA inspector and myself observed the passenger to be well orientated, emotionally stable, and completely rational when we interviewed him.

The passenger occupied the right cockpit seat. He was an FAA flight service station specialist, and a rather experienced single and multi-engine pilot. He gave a clear and concise account of his observations when we interviewed him while he was hospitalized after the accident. Being in related fields and having basically the same employer, we had much in common and enjoyed excellent rapport during our discussions. The primary bone of contention the attorney was gnawing on was related to my refusal to allow this passenger to alter his statement. In fairness to the passenger, the matter was never discussed with him. Therefore, we did not know whether he wanted to modify or alter his statements in any manner whatsoever.

The attorney submitted a rough flight path diagram that was purportedly drawn by the passenger. It contained notations of specific events occurring as the flight progressed. The diagram showed the pilot turning the fuel supply to the right engine OFF while executing the left turn back to the field. My refusal to even consider the matter was related to the below listed excerpts from the transcription of our tape-recorded interview with the passenger:

AUTHOR - - - - - "After he took off, did he reach down for those fuel valves?"

PASSENGER - - - "No, I don't think he did but I wouldn't swear he didn't. Why, were they in the OFF position or something?"

AUTHOR - - - - - "The right one is."

PASSENGER - - - "Well, the right engine was our most power."

FAA INSPECTOR - "Didn't you say it was surging, Jeff?"

PASSENGER - - - "Yes sir."

FAA INSPECTOR - "When that thing started losing power, Jeff, and started settling, your pilot instinct didn't make you look at those gauges to see what they were developing?"

PASSENGER - - - "I was looking out the window. I was looking for a place to

land. I knew he could fly the airplane better than I could because I have never flown a Navajo."

By way of explanation, the right engine fuel selector was found in the OFF position. The left engine fuel selector, that is directly adjacent to the right one, was found distorted to a position beyond its normal range of travel. The distortion of the left selector was in the same direction the right selector would be moved to turn it OFF. So, putting it in perspective, on the one hand we have the passenger stating he didn't think the pilot turned the fuel valves OFF and the right engine was developing the most power. On the other hand, we found the right engine fuel valve in the OFF position. It's not unusual to encounter such incongruities during accident investigations and the only rational way to resolve the issues is by a thorough analysis of all the pertinent evidence. Through this process, we didn't consider there was sufficient evidence to find that pilot factors were involved in the accident.

This occurrence makes a good test for deliberations about the probable cause. After all the evidence has been gathered, anyone should be capable of arriving at logical conclusions as to the factors involved. It's not unusual for there to be differences of opinion, but most individuals are generally in the same ballpark. In this instance, the attorney was trying to show some liability for the accident on the part of the operator. He broadcast the underlying rationale behind his persistence when he stated in one of his letters: "I am keenly interested inasmuch as I represent the estate of a deceased passenger with whom the insurance company does not choose to settle except for peanuts." Fortunately, the Safety Board does not have such a vested interest in the outcome and is able to weigh the evidence with complete impartiality.

FACE TO FACE WITH THE ENEMY

People cause most accidents, regardless of whether they occur in the home, on the job, on the highway, in the air, or wherever. Insofar as accidents are concerned, it can be stated, "We have met the enemy and it is us." A study of statistics on file with the National Transportation Safety Board will show 80 percent is a good ballpark figure for aircraft accidents that are pilot-caused. Of the remaining 20 percent, where some failure or malfunction of an aircraft system or component may have occurred, some pilot factors may also be involved.

For instance, engine failure, even on a single-engine airplane, does not necessarily cause an accident. Understand, no attempt is being made to infer that the engine failure didn't contribute to the accident. Oftentimes, however, it's just one happening in a sequence of events that terminates in an accident. The engine failure, which could have been caused by fuel starvation due to mismanagement of the fuel system by the pilot, may have occurred over an airport or other suitable emergency landing site. However, if the pilot, while attempting to cope with the engine-out emergency, loses control because he fails to maintain a safe airspeed, an accident may occur. This example is not used as a theoretical situation. Such circumstances are involved in a number of accidents.

- -

TOMBSTONE FACTORS

Factors are described as the circumstances, conditions, etc., that bring about a result. Tombstone factors relate to conditions the general public may not be fully aware of that have a major impact on safety, aviation or otherwise. They relate to circumstances where there is a problem area, and responsible officials are aware that an unsafe condition exists. However, nothing is done because, as yet, there have been no fatalities. After the problem causes a fatal accident, the tombstones become the catalyst that prods authorities to initiate actions designed to prevent similar occurrences.

For instance, birds present a hazard to aircraft operations, especially jet-powered models. Trash and garbage dumps attract birds; therefore, sanitary landfills are a hazard to aviation, and this fact is readily recognized by aviation authorities. However, trash dumps are still built in the near vicinity of airports and sometimes right on airport property. Fatal accidents have occurred at these locations because of bird strikes or bird ingestion in engine air intakes. It's amazing how fast governing bodies, boards, and commissions can find alternate sites to dispose of their rubbish when petitioners for change have some tombstones to line up before them. In the safety business, this is referred to as Tombstone Safety Legislation.

One such occurrence involved seven fatalities on a twin-engine corporate jet that crashed on takeoff. The takeoff path was over a residential area containing numerous apartment complexes, shopping centers, and busy thoroughfares. Immediately after takeoff, the airplane struck a flock of birds. This caused the loss of all thrust from one engine and about 60 percent from the other. The pilots were left with insufficient thrust to maintain level flight. The tower controller notified the pilot of smoke trailing from the engines after the flight lifted off the runway. The pilot replied they had hit some birds. The controller cleared the flight to land on any runway; but the pilot replied in a calm, clear voice, "We won't be able to make it."

A suitable emergency landing site was not available. The airplane initially collided with the roof of a three-story apartment and came to rest in a wooded ravine beside a busy highway. Three automobiles were destroyed in the postcrash fire, and a person on the ground sustained serious burn injuries.

Following an aircraft accident, local agencies respond immediately and extinguish any fires while rendering all possible assistance to survivors. The removal of fatally injured occupants is under the jurisdiction of the local medical examiner. Thereafter, authorities are requested to secure the crash site until Safety Board personnel arrive. In this instance, the accident occurred shortly after 10 a.m., and I arrived at the scene in the middle of the afternoon after taking a commercial flight from Miami, Florida.

When such tragic accidents occur in populated areas, things are always hectic. Much time is consumed conducting the necessary liaison with local authorities while otherwise organizing the investigation. The Investigator-in-Charge literally needs to be about ten places at once. A survey of the scene and discussion of the known facts provides the basis for an initial determination as to the scope and magnitude of the investigation. After taking actions to assure the preservation of the wreckage, it must be determined what technical assistance, if any, is going to be required. With so much to do, it was well after dark before I visited the airport control tower.

Only the controllers were on duty. While in the tower, one made a remark about the problems they were experiencing since the trash dump was put on the other side of the runway. When asked how far from the other side, he said it was on the airport, just on the other side of the runway. My spontaneous reply was, "For crying out loud, what's going on here, I thought we learned from the Lockheed Electra accident in Boston, Massachusetts, that locating trash and garbage dumps on, or in the immediate vicinity, of airports, is an absolute no-no."

The former military airport had been returned to the county for use as a civil airport. Since the county assumed control, six federal grants had been approved for the airport. Upon return of the airport to the county, and upon approval of each federal grant, the county gave assurances it would not permit any activity on the airport that would jeopardize aircraft operations. Additionally, the county agreed to restrict use of adjacent land to activities and purposes compatible with normal airport operations. The county disregarded its commitments, and started the sanitary landfill operation two years after assuming operational control of the airport.

During the investigation, large flocks of birds were observed flying over the airport, with birds numbering in the thousands swarming over the dump area. Prior to the accident, there had been much correspondence between the FAA and local authorities concerning the dump. Obviously, the FAA wanted the dump to be closed; however, the foot-dragging tactics by county authorities resulted in no concrete actions being taken.

Everyone interviewed who had knowledge of the situation stated they knew the birds attracted by the trash and garbage dump presented a hazard that would eventually cause an accident. Several months after the accident, it was learned that the Board of County Commissioners had some heated discussions concerning the municipal dump on the airport. Reportedly, one member was of the opinion that the dump was more important to the county than the airport.

All these carryings-ons graphically illustrate the tombstone aspects of the legislative actions taken after the accident. There is little doubt it took those seven tombstones of the accident victims, plus the threat of a cutoff of additional federal funds for the airport, to accomplish the required action.

Several years after the accident, I visited the airport manager while conducting another investigation in his locale. During our discussions, he mentioned that the aftermath of the earlier accident included the following:

> **The investigation you did involving the trash dump did more good than you could ever imagine. I've received numerous calls from airport managers all over the country telling me local authorities were getting ready to build a trash dump or sanitary landfill in the immediate vicinity of their airports. I advised them to protest, argue, shout, scream, and do anything else that wouldn't get them fired, to stop such actions. Many of them called back to tell me the tragic circumstances reported in the Safety Board's report, the recommendations made following the investigation, plus the threat of a cutoff of more federal funds, resulted in the dumps being built elsewhere.**

Before reviewing the Safety Board's probable cause determination for the accident, there's an issue relevant to airport management authority that might require clarification. At many locations, the manager is not the final authority for many decisions at his airport. This is particularly true at county and municipal airports when managers are employed by locally elected boards, commissions, etc. Oftentimes, their status requires that they go along with decisions, policies, and procedures established by these officials who may have no prior background or experience in aviation matters. Such was the case in this instance, and it was evident that the airport manager was deeply concerned that his efforts to remove the landfill from the vicinity of the airport had been unsuccessful.

The tombstone aspects of this investigation were exemplified in the Safety Board's probable cause determination:

> **"The National Transportation Safety Board determines that the probable cause of this accident was the loss of engine thrust during takeoff due to ingestion of birds by the engines, resulting in loss of control of the airplane. The Federal Aviation Administration and the Airport Authority were aware of the bird hazard at the airport; however, contrary to previous commitments, the airport management did not take positive action to remove the bird hazard from the airport environment."**

IVORY SOAP FACTORS

This term, and the implied 99 44/100 percent, has been plagiarized by investigators for occurrences wherein they are reasonably sure of the pertinent factors involved even before commencing to collect the relevant facts, circumstances, and conditions of an accident. For example, when both or all engines on multi-engine aircraft fail almost simultaneously, a fuel problem is indicated: either fuel exhaustion, meaning there is no fuel on the airplane, or fuel starvation, meaning there is fuel on the aircraft that is not available to the engines because of improperly positioned tank selector valves, or mismanagement of other fuel system components by the pilot or flightcrew. While fuel problems are the norm, investigators still have to conduct thorough inquiries because other factors, such as icing, failed components, etc., might have been involved.

In one accident, a small single-engine airplane was damaged but there were no injuries to persons on board. The pilot/owner stated he experienced complete failure of the engine. Investigation showed one fuel tank was full and the other completely empty. The pilot was advised that it looked as though his engine problems were related to fuel starvation brought about by his failure to switch tanks. However, he persisted with his story. The pilot was informed we would make arrangements to disassemble the engine. The engine had sustained very little damage, and the pilot wanted to know who was going to put it back together. He was advised we had authority to examine the wreckage to the extent required to complete our investigation. At the conclusion of the engine teardown, we would give him our "Release of Aircraft Wreckage Form." Thereafter, it would be his responsibility to return the airplane and engine to service. After pondering our response a few minutes, he asked if he could revise his statement to show he had failed to select another tank after running out of fuel. His request was granted, and we terminated the on-scene phase of the investigation.

As exemplified below, the less experienced pilots aren't the only ones guilty of mismanaging the fuel systems:

a. Mismanagement of fuel system occurred on the National Airlines Boeing 727 flight that landed in Jacksonville, Florida, with only one engine operating while on a flight from Miami, Florida, to New York.

b. Ten persons died when a four-engine United Airlines jetliner carrying 181 passengers crashed following fuel exhaustion to all engines while the crew was attempting to land at Portland, Oregon. The crew received an unsafe landing gear indication while preparing to land. Thereafter, the flight remained overhead more than an hour while they attempted to cope with the emergency.

The Safety Board's probable cause determination cited, "The failure of the captain to monitor properly the aircraft's fuel state and to properly respond to the low fuel state and to the other crewmember's advisories regarding the fuel state."

c. An Avianca Airlines (Colombia, South America) Boeing 707 crashed following fuel exhaustion to all four engines while attempting to land at the John F. Kennedy International Airport in New York after a flight from Bogota. Seventy-three of the 158 persons on board were killed. In its probable cause determination, the Safety Board cited, "the failure of the flightcrew to adequately manage the airplane's fuel load, and their failure to communicate an emergency fuel situation to air traffic control before fuel exhaustion occurred."

d. A Boeing 767 flight almost had to ditch in the Pacific Ocean shortly after takeoff from Los Angeles, California when the flightcrew inadvertently mismanaged the fuel controls.

Ivory soap occurrences are relatively easy to handle, except on those occasions when the pilot or flightcrew are uncooperative. In those instances, investigators must keep digging until they determine the factors involved, which, on occasion, have been found to be in conflict with statements made by the pilot or flightcrew.

PILOT FACTORS

Experience has proven, and statistics clearly show, pilot factors are the predominating cause of aircraft accidents. But understand, we're not attempting to blame everything on the pilots. There are limits to what can be expected from human performance. Mistakes are certainly not abnormal and they are going to continue to happen. If this book related to automotive safety, this chapter would be titled "DRIVER FACTORS;" or if it related to accidents in the home, it would reference "HOMEOWNER FACTORS;" because, as previously stated, people cause accidents.

Scoldings, or other adverse actions against guilty pilots, accomplishes nothing in the broad safety arena. Experience has proven that a more appropriate answer to accidents is better design of the machine or other product coupled with more complete and comprehensive training of the operators. Often, this is defined as "The Man/Machine Relationship." It sometimes appears that airplanes are manufactured with occupant comfort in mind as opposed to overall safety. The fact that the Safety Board once conducted a special study entitled, "Design Induced Pilot Error," substantiates that, on some occasions, there is a better way. Once people get deeply involved in safety in any field of human activity, it spills over into everything they see and do. We just moved into a subdivision where construction is ongoing. It is not unusual to observe workmen engaging in unsafe activities, or taking all manner of unnecessary risks. So, what are the factors that cause pilots to have accidents? Actually, there are many, but let's look at a few that occur with amazing regularity:

OVERCONFIDENCE - This factor is involved in many weather-related accidents where noninstrument-rated pilots attempt visual flight in instrument weather. They might take off in instrument conditions from an uncontrolled airport (that's one without a control tower) or continue visual flight into nasty weather they encounter en route.

It's pretty hard to confuse pilots as to which end is up as long as they can see the ground or horizon. But take those away, and pilots who haven't received the requisite training to fly their airplane by sole reference to the flight instruments will lose control on almost every occasion.

Without going too deeply into aviation physiology, we all need to understand that our basic sense of balance is located in the middle ear. As long as we stand in a fixed position on the good earth with our eyes open, we're not likely to become disoriented unless we've been overly indulging in the corn squeezings. Remember when we were kids, they'd blindfold us, then turn us round and around a few times. They were just confusing our middle ear as to which end was up. Even after removing the blindfold, everyone laughed as we staggered about until the middle ear mechanism righted itself. It's sometimes called vertigo, or spatial disorientation, or space myopia, but whatever, it basically means we don't know which end is up.

Weather is one of the predominating factors in many fatal occurrences. It's said everyone talks about the weather but nobody does anything about it. Well, pilots are in a position to do something about the weather, because they can elect to stay on the ground. My investigations included many occurrences where novice pilots were attempting visual flight in weather conditions professional pilots would avoid unless on an instrument flight plan.

On one occasion, four persons died when their airplane crashed in pea-soup fog on takeoff from a western North Carolina airport for a flight to the coast. When they planned the flight earlier in the week, the low-time, non-instrument-rated pilot, should have said, "We'll go if the weather's okay." However, once everyone arrived at the airport, keenly anticipating a pleasant weekend on the beach, the flight apparently became more urgent. This might have caused the pilot to feel some pressure to attempt the takeoff. If you're ever a passenger on a light plane where the weather is marginal and you're not sure of the pilot's qualifications, don't be afraid to ask what certificates he holds and how much experience he has. If you have any doubts, scratch your name from the manifest. You may be doing a whole world of good for your longevity.

Another accident involved a pilot who crashed while on a visual flight. A stationary front was located on the Florida/Georgia line. On many occasions, frontal activity has a tendency to stagnate along that line. The pilot, with his family on board, took off from an airport south of the frontal activity where visual weather was prevailing. He was required to taxi around a long line of airplanes awaiting instrument flight clearances through the weather. However, since the weather at the departure airport was above visual minimums, the pilot violated no regulations when he took off.

The flight proceeded normally until encountering the weather. After penetrating the frontal activity, the pilot, lacking the requisite instrument flight training, became disoriented, lost control, and the airplane entered an inadvertent spiraling dive, sometimes referred to as a graveyard spiral. Witnesses said they heard it in a "power dive" and reported an explosion before it crashed. There was no explosion as such, but the airplane did sustain an in-flight structural breakup, (meaning the wings and/or tail came off,) and that always sounds like an explosion. Invariably, such structural breakups occur as the pilot manipulates the controls in an effort to recover. Investigators would say the breakups were caused by pilot-induced aerodynamic loads in excess of the design limitations.

What actually happens is the pilot, upon coming out of the clouds, regains visual reference with the ground and is immediately reoriented as to which way up is. He instinctively pulls back hard on the controls, and an instantaneous breakup of the aircraft structure occurs. The airplane comes apart because the pilot exceeds the maneuvering speed. This is a relatively low maximum airspeed where an abrupt movement of the flight controls can be made without exceeding the authorized structural aerodynamic limits for the airplane. However, when a pilot pops out of the base of clouds, speeding towards "terra firma" with the airspeed needle wound up in the eight to twelve o'clock range on the dial, human nature almost dictates that a rapid pullback on the controls be made in an attempt to raise the nose. And that's when the "explosion" occurs.

There are numerous occurrences where unqualified, noninstrument-rated pilots attempt instrument landing approaches in weather that might tax some professional pilots' abilities. A noninstrument-rated pilot attempted a night approach in nasty weather showing a ragged 300-foot ceiling. His spouse and another couple were on board, and his mother was at the airport with his children whom he intended to enplane before continuing on home. He executed the missed approach procedure after missing the first approach. The airplane was not sufficiently aligned with the runway to complete a landing on the second approach. Instead of pulling up for another go-around, as required, he crashed while attempting to circle for landing underneath the overcast, killing all four people on board.

Investigators refer to such occurrences as "get-home-itis" accidents. The dictionary might define get-home-itis as, "an abnormal state or condition, excess tendency, obsession, etc., to proceed to one's domicile without due regard for one's own safety or those in the immediate vicinity." The frequent happening of similar accidents shows get-home-itis can be a powerful force. Investigations show many pilots will attempt to stretch their gasoline supplies, exceed their flying abilities, penetrate all manner of weather phenomena, or do whatever else might be required, to return home on their preplanned schedule.

Similar accidents often involve well qualified pilots who are willing to stretch their luck by busting the weather minimums, as it's called in the trade, while attempting to land. In another occurrence, a pilot possessing an airline transport certificate tried to circle and land when the weather was below the prescribed minimums. Witnesses at the airport heard the airplane circling overhead at low altitude before it crashed on the airport, killing both occupants. Understand, attempts to circle and land in such instances are not left to the discretion of the pilot. Pilots are in violation of the FAA's rules for instrument flight when they fail to execute the prescribed missed approach procedure if, upon reaching the published minimums, they don't have visual reference with the runway environment.

Another fatal accident involved an experienced pilot who was the operator of an airport not equipped with any instrument landing aids. The accident occurred while he was attempting to land at night under restricted ceiling and visibility conditions. When asked why the pilot was trying to land in such nasty weather, witnesses said he was using a procedure he had devised and used in similar weather many times in the past. The only thing to be learned from this, and similar occurrences, is that so long as we have pilots willing to take such chances when they have full knowledge of the risks involved, job security for aviation accident investigators is assured.

LACK OF FAMILIARITY WITH THE AIRPLANE This factor causes many accidents annually. Actually, they are so numerous one might think there is a flaw in the FAA regulations that grant authority for appropriately rated pilots to fly small, single and multi-engine airplanes. Possession of an automotive driver's license authorizes the holder to operate anything from a Volkswagen Bug to a stretch Lincoln Continental. Likewise, a pilot holding a single and multi-engine airplane rating is authorized to fly most of the small, single, and multi-engine airplanes on the market. We all recognize that there are great similarities in automobiles regardless of where they were manufactured. However, the same is certainly not true of airplanes. There are vast differences in the engines, fuel and other systems, operating speeds, emergency procedures, and many other areas vital to their safe operation.

Admittedly, there is an assumption by the FAA that pilots, in compliance with the regulations, will acquire the requisite knowledge about a particular model before they attempt to fly it. However, accident statistics will show there were numerous occasions when they didn't. And we're not referring only to inexperienced pilots, because some, with seemingly impeccable qualifications, have crashed because they didn't have the required knowledge about their airplane. There is an old saying: "familiarity breeds contempt," but it certainly doesn't apply to a pilot and his airplane. It's doubtful whether there has ever been an accident caused by the pilot having too much knowledge about his flying machine.

When discussing pilot qualifications and knowledge of their airplanes, it's well to look at the military services. Aviation safety is a primary concern in all but wartime, where mission accomplishment becomes a consideration. The military pilot must show aeronautical aptitude before ever going to flight school. Once in school, he or she receives the finest and most comprehensive training available. Those not progressing normally, or failing to show the requisite aptitude for flying, are washed out of the programs. This is contrary to much civilian flight training where students are rarely washed out, unless, of course, they run short of funds to pay for their training.

There is a requirement by the Pentagon for manufacturers to provide detailed operating instructions for their airplanes. Military commanders insure that their pilots know and thoroughly understand these instructions. One of the surest ways to injure a military career is for a squadron, group, or wing commander to have an unacceptable aircraft accident rate.

The military computes its aircraft accident rates by the number of major accidents in a stipulated number of flight hours. After WW II, military aircraft prices skyrocketed, and the services realized that a significant improvement in the accident rate would be required just to keep from going broke buying replacements. Formal aviation safety officer schools were opened, and intensive training was instituted using flight simulators and other sophisticated training devices. New aviation safety officer billets were established at the squadron, group, and wing levels. Comprehensive standing operating procedures were promulgated, and in-depth knowledge and strict compliance with prescribed emergency procedures were stressed.

The fact that military commanders have authority to ground, or otherwise prevent pilots assigned to their units from flying, was a positive factor. The military aviation safety program produced astounding results. An eight to tenfold improvement in the accident rate was achieved in a relatively short period of time. When it's considered that some current military models cost in excess of a half billion dollars each, it's clearly evident that a significant improvement in the accident rate was mandatory.

COMPLACENCY - Complacency signifies contentment, self-satisfaction, or smugness. While it may be okay in some areas, it certainly doesn't contribute to a pilot's flying ability or to his or her total life span.

An old proverb says: "Those who know little, are confident in everything." Pilots don't become complacent intentionally. Instead, it's an outgrowth of their indifferent approach regarding knowledge of their airplanes, compliance with recent flight experience requirements prescribed by the FAA, overconfidence in their abilities, etc. In other words, they approach and study the various facets of aviation with about the same degree of enthusiasm they exhibit towards automobile driving. Pilots with lots of experience need to guard against becoming complacent. Otherwise, they can fall prey to the cliche, "the good die young." That is true, because all their experience can cause them to let their guard down. Conversely, inexperienced pilots, realizing they don't know everything, are prone to be more careful. A statement often heard by the investigator is, "He or she was one of our best pilots."

The speeds prevalent in aviation today have greatly reduced the time required to travel between two points. While that's true, the accident investigator has learned it also applies to points between this world and the next. If pilots are doing nothing to increase their knowledge of aviation, they are probably regressing somewhat in knowledge they once possessed. Unlike the military pilot, general aviation pilots don't have anyone to closely monitor their aviation activities. In reality, they're operating in a self-policing environment. The only proper attitude is to approach aviation with self-discipline, determination, restraint, and control, while always striving to increase knowledge about the many facets of aviation and gain flight experience. With attitudes such as these, and a firm commitment never to gamble with marginal weather, a pilot will have a jump start on someday being referred to as, "An Old Pilot."

A telephone call was once received from a reporter who wanted to discuss "The structural breakup problems we're experiencing with the Beech Bonanza." He called me because I had an ongoing investigation involving a Bonanza that experienced an "explosion" and in-flight breakup after coming out of the bottom of a severe thunderstorm. My explanation that the problems were not associated with the construction of the Bonanza, but rather with the qualifications of the pilots flying them, caused him to become almost vitriolic.

He refused to accept the fact that numerous studies had shown the Bonanzas exceeded the airworthiness criteria established by the FAA, plus the fact that it took no extraordinary pilot ability to fly them. It was explained that many Bonanzas are equipped with engines powerful enough to obtain optimum speeds, and, since the airplane is so aerodynamically clean, it whistles through the air without a lot of protrusions to slow it down. This lets the Bonanza build speed rapidly when in a dive, regardless of whether the dive is intentional, or the result of a loss of control by the pilot.

Shortly thereafter, one of the Safety Board's engineers, whom I had worked with previously and respected very much, called from Washington. He told me the reporter had just called him and the essence of their conversation was my refusal to go along with the reporter's point of view. I had far more flying experience than our engineer. After a long discussion of the relevant facts and conditions, he indicated general agreement with my position.

Some time later I watched a television show concerning in-flight breakups of the Bonanza. This was one of those affairs where the media use bits and pieces of recorded conversations with various parties they have interviewed. It was utterly amazing how they could substantiate, to some degree, the facts required to justify their conclusions by using a word or phrase from one interview, here, followed by a word or phrase from another interview, there. As the saying goes, "The show probably played well in Peoria," but few in the audience had the requisite experience and background to fully comprehend the full measure of the opinions expressed.

There's one other issue to be covered before we leave this subject. We all realize and cherish the fact that we are living in a free country. Whenever our federal government establishes regulations, standards, etc., they are always expressed in terms relating to the minimums required for compliance. This is true of the airworthiness standards established by the FAA for the manufacture of various categories of airplanes. As long as a particular airplane meets the established criteria, and does not require unique or unusual pilot skills, it is in compliance with the regulations. There are no requirements for the FAA to dictate the design of, or impose limitations on, a particular airplane to assure complete safety despite flagrant violations of flying regulations. Neither is the FAA required to make amends for instances of extremely poor airmanship. This is a vital point that should be borne in mind, because it is applicable to many of the occurrences we will be discussing.

ALCOHOL AND DRUGS - The use of these products by pilots is probably more prevalent than most folks realize. Maybe I'm just naive, but I was flabbergasted on several occasions when the next-of-kin replied: "Oh, we knew that," when they were informed the toxicological studies showed drug, or both alcohol and drug use, by a pilot involved in an accident.

Use of, and the effects of alcohol and drugs, fall in the human factors arena. As with the automobile, the use of these substances is involved in a great number of accidents. We've probably all heard that a 0.10, or higher, blood-alcohol level is used to show whether an individual is under the influence. Such a level puts people in the euphoric stage where they feel happy, buoyant, vigorous, talkative, and slightly light-headed. An individual with a lower concentration may not be legally under the influence; however, tests have shown concentrations less than 0.10 are enough to affect a pilot's ability and judgment.

Higher altitudes tend to magnify the effects of alcoholic consumption. Drinking three highballs in Denver, Colorado, produces a more intoxicating effect than having those same three highballs in Miami, Florida. The lower atmospheric pressure at Denver's 5,000-feet altitude equates to less oxygen being available in a given volume of air. Additionally, there were occurrences where toxicological studies showed pilots were under the effects of both alcohol and drugs. Upon asking the toxicologist what the cumulative effects would be, he replied, "Your guess is as good as mine. There's no data available we can use to quantify it."

We may as well go to the top of the class to discuss alcohol-related occurrences. One of my investigations involved a pilot with one of the highest blood-alcohol concentrations ever documented, a 0.593 level. The blood-alcohol effects chart goes through the various concentrations from euphoria, excitement, confusion, stupor, coma, with death occurring at concentrations above 0.45 due to respiratory paralysis. However, there is a clarifying remark that some individuals have been known to survive levels as high as 0.60.

The pilot was reportedly having some drinks with friends in a bar when he declared he was going to fly by in his airplane. The passenger, who willingly accompanied the pilot, had also been in the bar. Studies showed his blood alcohol concentration was in the "excitement" range. It was reasonable to assume he wouldn't have gone along if the pilot's demeanor, gait, and manner of speech had shown him to be under heavy alcoholic influence. A short time later, their friends in the bar heard the airplane overhead and went outside to observe the pilot making low passes and steep pullups. On the final pullup, the nose fell through and the airplane dove into the ground, killing both occupants.

While documenting the wreckage at the crash scene, a gentleman arrived and stated he wanted to talk to whoever was in charge. After introducing myself as the Investigator-in-Charge, we moved out of earshot of others present. He stated the pilot was drunk. When asked how he knew, he said the pilot hadn't drawn a sober breath for 30 years. He went on to say he was the pilot's neighbor, and he and other neighbors had permission from the pilot's spouse to park their cars around the airplane, or to remove the propeller, to keep him from flying when it was obvious he was too much under the influence. He was advised the witnesses at the bar stated the pilot had been drinking, and the medical examiner reported a strong odor of alcohol during the post-mortem studies. I thanked him for his information and concerns, told him toxicological studies would be conducted, and informed him positive findings would constitute "prima facie" evidence of alcoholic involvement.

Blood and urine samples are the primary specimen toxicologists use for analysis of alcohol, drugs, carbon monoxide, etc. The specimen collected for this accident were shipped to a government laboratory for analysis. The senior toxicologist called after the studies were completed to say they couldn't believe their findings. Pursuant to his request, additional specimen were forwarded that confirmed the validity of the initial studies. It was learned from discussions with the toxicologist and other medically orientated personnel that it's possible for an individual to build up a significant tolerance to the effects of alcohol through heavy consumption over an extended period of time.

Another alcohol-related occurrence involved a light, single-engine airplane that crashed while buzzing a tavern in which the occupants had been drinking. The pilot and passenger arrived at the tavern about 6:00 p.m. About 9:30 p.m., a witness heard the following conversation:

> **Pilot, - "If you've got the guts to ride with me, I've got the guts to fly it."**
>
> **Passenger, - "Well, if you've got the guts to fly it, I've sure got the guts to ride with you."**

They departed immediately and about 10:00 p.m., a patron in the tavern heard the airplane overhead. He went outside and observed the airplane approaching at a low height above the terrain. It flew into the top of a tree and collided with a chimney and antenna on the roof of the tavern before crashing on an asphalt roadway. The airplane burst into flames and the fire was so intense that would-be rescuers were unable to render assistance.

- -

P-51 FACTORS

This is a method a few pilots resort to in an effort to show flight experience in their logbooks. However, to find out who's being fooled, all they need do is look in a mirror.

The practice relates to instances where pilots enter flying time in their logbooks for flights that never took place. In other words, they were "padding" their total flight experience. The term "P-51", refers to the use of a "Parker 51" fountain pen to make the logbook entries. The practice is not all that uncommon. It was uncovered during several investigations when significant discrepancies were found in the times entered in the pilot's logbook in a particular airplane, and the actual time that airplane had been operated, as recorded on the hourmeter installed on the cockpit instrument panel. Although such padding might not be the direct cause of an accident, the pilot's total flight experience is oftentimes considered when analyzing the evidence collected during an investigation.

WIND SHEAR, A DEADLY FORCE

"An approach which places an airplane in or near a thunderstorm at low altitude is hazardous. The wind conditions which might exist can place an airplane in a position from which recovery is impossible—even if both the pilot and the airplane perform perfectly." The above statement was excerpted from the Safety Board's report of the Eastern Air Lines Boeing 727 landing accident at the John F. Kennedy International Airport in New York in June 1975. In August 1985, a Delta Air Lines Lockheed L-1011 airplane crashed while approaching through thunderstorm activity to land at Dallas, Texas. In both occurrences the Safety Board cited the flightcrew's decisions to continue their approaches through weather known to be hazardous as contributory factors to the accidents.

The low-level wind condition causing the greatest hazard to airplanes is called wind shear. During a 15-year period beginning in the 1970's the Safety Board identified low-level wind shear as a cause or contributing factor in 16 accidents involving transport airplanes. Nine of those accidents were nonfatal, but the other seven resulted in 574 fatalities.

Just what is wind shear? According to the FAA: "Wind shear is best described as a change in wind direction and/or speed in a very short distance in the atmosphere. Under certain conditions, the atmosphere is capable of producing some dramatic shears very close to the ground; for example, wind direction changes of 180 degrees and speed changes of 50 knots or more within 200 feet of the ground have been observed." In cases where the wind speed changes faster than the aircraft mass can be accelerated or decelerated, wind shear can have a devastating effect on an airplane's performance capabilities.

The most prominent meteorological phenomena causing significant low level wind shear are thunderstorms and certain frontal systems at or near airports. The winds around a thunderstorm are complex. Wind shear can be found on all sides of a thunderstorm cell and in the downdraft directly under the cell. The downdraft, downburst, or microburst, as it is often called, is part of the evaporation-condensation process which produces cumulonimbus clouds, heavy rainshowers, and thunderstorms. Low-level air, heated by the ground, rises and is replaced by cold air descending from above. As the column of colder air strikes the ground, it fans out in all directions.

The condition can be simulated by directing a heavy stream of water vertically onto a paved surface. Upon striking the ground, the stream of water bounces off the pavement and fans out in all directions. An airplane flying into the descending column of air typically experiences an increasing headwind, followed by a downdraft and then a tailwind in rapid succession. During its investigation of the Delta Air Lines Lockheed L-1011 accident at Dallas, the Safety Board found the airplane's path through the downburst resulted in an airspeed loss of 44 knots in only 10 seconds while the downdraft velocity was increasing to 1,800 feet per minute.

An airplane encountering such significant downdrafts when operating at low heights above the terrain can literally be driven into the ground. It should be realized that an airplane always moves with the surrounding air in which it is operating. That's why it is necessary to crab (turn into the wind) to counter the sideways motion over the ground when a crosswind exists. In cases where an airplane is descending at a rate of 500 feet per minute, and the block of air in which it is operating suddenly starts descending at a rate of 1,500 feet per minute, the actual descent rate for the airplane is 2,000 feet per minute.

The prevailing weather conditions in the accidents cited above were known to both the flightcrews and the air traffic controllers. Controllers provide the pilots of landing aircraft with the existing weather conditions. However, the final authority for the safe operation of any flight rests with the pilot-in-command. This precept should be kept in focus; it comes into play for accidents discussed in other chapters.

The Safety Board's report for the New York accident stated: "Pilots must exercise more independent judgments when they are confronted with severe weather conditions in the terminal areas. They must recognize that the conditions within, under, or near rapidly developing and maturing thunderstorms are dynamic and can change significantly within a short distance, or within a short time, or both. In particular, they must recognize and avoid low-altitude hazards associated with thunderstorms along or near the approach path." However, as we shall learn, there are occasions where other factors come into play that pressure the pilot-in-command to make an earnest effort to land at the flight's intended destination.

Let's review aspects of the investigation into the Eastern Air Lines Lockheed L-1011 trijet accident in the Everglades Swamp near Miami, Florida. While the flight crewmembers were coping with an unsafe nose landing gear light, the readout of the cockpit voice recorder showed the copilot saying, "Always something, we could have made schedule." This "making schedule" thing carries lots of weight in the airline industry. Air carriers compete against each other; the government maintains statistics and reports the winners and also-rans periodically, and, without obvious pressure from management, it's something flightcrews are always striving to achieve.

"Making schedule" was alluded to in the Safety Board's report of the Eastern Air Lines wind shear accident in New York. The report stated: "When operating at capacity, the air traffic system in a high density terminal area tends to resist changes that disrupt or further delay the orderly flow of traffic.... Consequently, controllers and pilots tend to keep the traffic moving, particularly the arrival traffic because delays involve the consumption of fuel and tardy or missed connections with other flights, which could lead to further complications. As the weather conditions worsen, the system becomes even less flexible."

As a result of its investigation of the New York accident, the Safety Board made 14 recommendations, all related either directly or indirectly to the weather. They called for better wind shear detection devices, better training for flightcrew members and air traffic control personnel, research to develop better equipment, etc. Funding provided to conduct extensive studies of the wind shear phenomena has produced significant benefits. Air carrier pilot training has been updated and procedures have been implemented to reduce the hazards in wind shear encounters. Additionally, FAA regulations have been revised to require the installation of wind shear detection equipment on jet transport models in the early 1990s.

Despite these advances and predicted future benefits from the ongoing studies, it must be recognized that wind shears are dynamic and transitory. Weather forecasters fully understand that we will never be able to forecast accurately the precise location where, and moment when, all significant wind shears will occur. Therefore, since the most violent wind shear activity is produced by the convective movement of air around thunderstorms and towering cumulonimbus buildups, pilots should be wary when flying under or in close proximity to such activity during the final phases of their landing approaches, and their initial climb-outs after taking off.

The National Weather Service utilizes its weather surveillance radar to classify thunderstorms and other buildups containing enough moisture to provide a weather radar echo. It is important to understand that weather radar equipment only depicts water in the atmosphere. Currently, there is no radar equipment that will depict all cloud formations nor is the present equipment capable of accurately detecting turbulence. However, the existing equipment permits weather radar observers to determine objectively the intensities of the radar weather echoes. Based on this capability, the weather service has classified six levels of echo intensity and provided a correlation between those intensities to categorize the degree of turbulence in a particular buildup. The turbulence in level four, five, and six returns are labeled VERY STRONG, INTENSE, and EXTREME, respectively.

The pilot must make a judgment call regarding appropriate actions to be taken after weighing all known factors when conditions are such that wind shear may be encountered. Wind shear is not something to be avoided at all times, but rather to be assessed and avoided if severe. In this regard, radar echoes classified as level four, five, or six should alert pilots to take appropriate precautions.

When a hurricane is approaching our shores, the Director of the National Hurricane Center in Miami, Florida routinely appears on our TV screens. Hurricane Center spokesmen exhibit a wealth of knowledge while providing updates on the storm in language we all understand. They keep us advised of the point where the hurricane is likely to strike land. Their predictions are based on an analysis of the storm's prior route, and the manner in which the prevailing weather systems might affect its path. However, the Director always makes it clear we should stay tuned for further developments, because it is impossible to state with certainty whether the hurricane will weaken or strengthen significantly, or just where a particular storm might come ashore.

Similar conditions exist for thunderstorms and other weather phenomena. Significant improvements have been made that are materially assisting pilots in their decision-making process. Depending upon the landing speed of their airplanes and the length of runways in use, pilots can sometimes adjust their approach and landing speeds sufficiently to cope with rather significant wind shears. However, it has to be recognized that we mortal beings will never be able predict the exact paths storms will take. We have come to realize that only a small error in our predictions or forecasts can result in dangerous conditions for landing or departing flights.

It is probable that our Creator never meant for us to fully understand and comprehend every facet of his omniscient and omnipotent powers. Those who don't believe in the Creator philosophy may substitute the word Nature, because Nature hasn't, and never will, reveal all her secrets either. So we go back to the Safety Board's recommendations following its investigations of accidents attributed to wind shear. They can install sensors, Doppler radar devices, develop laser beam technology, install wind shear detection equipment around airports and on airplanes, or whatever; but we still won't know the exact magnitude of the downburst within a storm system, or the precise path the storm or weather phenomenon will take from its present location.

That leaves us with a future that is not too bright, since we are dealing with an imperfect forecasting system and fallible mankind. Accordingly, it is reasonable to assume we haven't seen our last air carrier accident resulting from encounters with wind shear. This is particularly true if operations are continued when well developed weather phenomena associated with significant wind shear are astride the final landing approach course, or the initial departure path, after taking off.

Landing accidents have always accounted for a high percentage of aircraft accidents. The wind shear aspects of the Eastern Air Lines Boeing 727 accident in New York received wide distribution. Additional publicity was generated when the Safety Board, FAA, air line companies, the Air Line Pilots Association, and others involved with aviation safety began showing renewed interest in wind shear as a factor in aircraft accidents. It didn't take long for light airplane pilots to latch onto the wind shear phenomenon as a crutch to lean on when reporting their landing miscues. The statement, "Encountered wind shear on final approach and crashed short of the runway," came into fairly common usage by pilots completing the accident report forms required by the Safety Board. To them, it made little difference whether the prevailing meteorological phenomena associated with wind shear were present.

But then the cycle ran full circle. A pilot was executing a landing approach to an airport with a short landing strip. In an effort to land close to the end, he failed to maintain flying speed. This caused the airplane to stall, and he crashed short of the runway threshold. In his account of the accident, the pilot reported he lined up with the runway and everything was fine until he crossed the boundary fence. Thereafter, "because of the severe calm," the airspeed dissipated rapidly and the accident occurred moments later.

It is recognized that a "becalmed" sailing vessel might be in some difficulty. However, we may well go through a whole host of February 29ths before we have another aircraft accident that will be attributed to "severe calm" wind conditions.

PEOPLE AND PLACES

One of the most interesting aspects of being an aircraft accident investigator is you never know where you will go next or whom you will meet. And I'm not convinced it's the places you go, as much as it is the folks you meet, that makes it exciting. Aircraft accidents can and do occur almost anywhere. When initially employed in the Safety Board's Miami, Florida office, our area of responsibility included Mississippi, Tennessee, North and South Carolina, Georgia, and Florida, plus the Bahama and Caribbean Islands. We responded with on-scene investigations to the more serious occurrences that usually involved fatalities. One thing I learned was the realization that the citizens of our country will do everything possible to aid the post-accident proceedings when fatalities are involved.

Investigators are issued books of government transportation requests they are authorized to sign. This was essential because we oftentimes responded to accidents from our homes during off-duty hours. Safety Board investigators are also issued credentials that authorize them to occupy the cockpit jump seat, an extra seat in the flightcrew compartment. We requested the jump seat when the cabin was full, but also used it as a training aid. When on the jump seat, we listened to all communications with the flight. Additionally, the necessary interaction with the flightcrew and opportunity to observe them in the performance of their duties kept us abreast of the current state of the art for airline operations. I saw some old friends and made many new ones while occupying the jump seat. It was indeed a rare occasion to ride with a flightcrew that did not extend the utmost cordiality.

We were also authorized to rent automobiles and small airplanes, hire stenographers and mechanics, contract with wreckage salvage operators, etc. However, there is a limit to the total expenditures an investigator is permitted to incur for a particular investigation, and the Safety Board has in place the necessary checks to ensure that such authorization is not abused for personal benefit.

- -

"MY GYRO'S BEGINNING TO SPIN"

An accident, in which the pilot and passenger died, occurred in Georgia. The rather well-qualified pilot was on an IFR (instrument flight rules) flight plan in instrument meteorological conditions. The occupants were returning to their Florida homes after a trip that included stops in South Dakota and Wisconsin.

While in level cruise flight, the pilot was advised by the controller that he appeared to be turning off course. The pilot replied: "I know, it looks like my gyro's beginning to spin." The circumstances showed the pilot failed to regain control, and the airplane continued descending until it crashed. Based on my experiences as a pilot and accident investigator, it is a rare occasion when a pilot simultaneously loses all the gyro-operated flight instruments.

The gyro-operated flight instruments present a near pictorial display of the attitude of the airplane in relation to the horizon. On occasions where the gyro-operated instruments fail, the pilot has other instruments that can be used to maintain control during actual instrument weather. The needle-ball indicator is used to control the bank angle, and a crosscheck of the altimeter, airspeed, and rate of climb indicators are used to control the position of the nose in relation to the horizon. In instrument flight schools this is called partial panel flying. It takes a much higher degree of instrument flight proficiency to maintain control during partial panel flight. That old Nemesis "complacency" oftentimes comes into play and, after obtaining their instrument certificates, many pilots fail to maintain the degree of proficiency required to fly their airplanes safely when confronted with partial panel flight.

The investigation of such occurrences is a lengthy process. Every precaution must be made to avoid destroying significant evidence. In this instance, our careful documentation of the wreckage and removal of the flight instruments for analysis at an FAA-certificated facility failed to provide evidence of the factors that caused the loss of control by the pilot. Failing to find the precise nature of the circumstances encountered by the pilot was always reason for concern. However, it is always satisfying to know a most thorough and complete investigation has been conducted.

Some of the victims' next of kin visited the crash scene. They collected the personal belongings, but when ready to leave, it was noticed they hadn't taken the cargo, consisting of some frozen food and a significant quantity of cheeses. Upon being reminded, they said they didn't want it. They were advised we were not authorized to dispose of it. They gave an affirmative reply when I asked if they had any objections if we let a black man who lived nearby take it. It was later learned the victims did some hunting in South Dakota, and the frozen food was pheasants.

The accident scene was wooded and access was down a country lane running by the recipient's home. The building was a weathered frame structure with a sun-faded coat of barn-red paint. It had a porch all across the front and down one side. Six to eight children were playing in the yard and climbing the porch railing when I initially had stopped to ask directions to the crash site. Upon departing from the accident scene several days later, the man who had taken the food was met coming the opposite way in his vintage pickup truck. We stopped alongside each other and I thanked him for his assistance. While chatting, it was learned he was the children's grandfather. When asked whether the frozen birds were pheasant, he replied: "Capt'n, I don't know if they's pheasant or what they is, but us and them chillens is eating away at 'em, and they's mighty good."

WELL KEPT SECRET

Cedar Key, Florida, consisting of a group of small islands off an isolated stretch of Florida's Gulf coast about 110 miles north of Tampa, may be one of Florida's best kept secrets. To get there, you turn left off U. S. Highway 19, the main highway between Clearwater and Tallahassee, and drive through 22 miles of nothing before arriving at the tiny fishing village. However, some have found flying into the island's airport is a better mode of travel and it has become a popular retreat where many light airplane pilots bring their families and friends for relaxation and an excellent seafood dinner. On most occasions, they arrive during daylight hours and depart the same evening.

A commercial flight was taken to Gainesville, Florida on my initial visit. After stopping at the county sheriff's office in Bronson, Florida, I drove on to Cedar Key in a rented automobile. The sheriff told me his deputy in Cedar Key would assist me in any manner possible. When asked how I'd locate his deputy, he replied: "Well, his name's Tiny Bell, and he's the biggest thing down there, so you shouldn't have too much trouble."

A sign alongside the highway into town told of the high school basketball team's victory in the State's 1A championship tournament several years earlier. Actually, it was quite an achievement because Cedar Key was the smallest school in the state. Even in 1991, the school's enrollment in grades kindergarten through high school was only 200 students, with the June graduating class consisting of 12 seniors. The quaint community had many old structures along its somewhat narrow streets, but a number of first-class seafood restaurants on the waterfront. The Florida Marine Patrol has a large contingent at Cedar Key and the Coast Guard has men assigned to care for the buoys, markers, etc., in the vicinity. Both contingents responded to our requests for assistance, and with so much water about, investigators from our office enlisted their help on many occasions.

Tiny and the FAA inspector from the St. Petersburg, Florida office were found at the crash site near the airport on the opposite side of the town. Tiny, who weighed about 350 pounds, told me I should obtain a room, because they were scarce that time of year. The FAA inspector had checked into the ancient hotel, and I drove back to reserve a room. The counter in the dimly lit lobby was unattended, but after a couple of minutes, a raspy female voice from a dark corner behind me brusquely asked: "What you want?" I expressed my need for a room for a couple of nights and she asked who I was. Upon learning I was a government employee in town to investigate the airplane accident, she said she had rented a room to another government man and it had two beds. Answering facetiously, I told her he snored. While still reclining on her couch, she replied: "You mean I have to get you another room just because he snores? Well all right, you go on back out there, and I'll have you a room when you get back."

Upon returning to the accident scene, Tiny was asked where they had found the charming innkeeper. He said: "Oh, I see you met Bessie. Well, Bessie's been over there insulting people for years, and most of them just can't wait to come back."

The rooms were spotlessly clean with comfortable mattresses, but the furnishings had an antique appearance, and the bathroom fixtures were from an earlier era. Over the years, the small stand-alone porcelain lavatory and four-legged tub had accumulated rust deposits below the faucets that no degree of scrubbing would ever remove. We had dinner at one of the waterfront restaurants the first night and, upon Tiny's suggestion, patronized the hotel restaurant the following evening. We found Bessie to be a most interesting and delightful hostess, once we got to know her, of course. Or maybe it was after she got to know us, but at any rate, I drove out of town several days later with a warm feeling in my heart for Cedar Key and everybody in it.

The airport in Cedar Key had one runway. On takeoffs to the south, airplanes flew over the town, but on takeoffs to the north, they flew over nothing insofar as a development or lighting on the ground was concerned. Night takeoffs to the south presented no problems because there was ample lighting on the ground to permit pilots to fly by visual references. However, on night takeoffs to the north, pilots were looking down into nothing but a seemingly black hole after passing the runway lights. Under such circumstances, unwary pilots may inadvertently fly their airplanes into the ground or water. It happens because it is almost impossible to determine your height above the water or terrain by visual cues when there is nothing but total darkness below. Lacking a more descriptive phrase, I borrowed one, and referred to it as the black hole phenomenon. Pilots flying under such conditions must crosscheck the altimeter, airspeed, and rate of climb indicators to assure they are climbing to, or maintaining, a safe altitude.

This black hole stuff we've been discussing is the primary reason most investigators assigned to the Miami Regional Office have, or eventually will, cover an occurrence in Cedar Key. I also investigated occurrences in Tampa, Florida, where a small airplane pilot inadvertently flew into the water while executing a night visual approach to a waterfront airport; and another at Summerton, South Carolina, where a pilot inadvertently flew his airplane into the water following a night takeoff from a waterfront airport.

Such accidents are not reserved solely for the inexperienced or unwary general aviation pilot. One of my investigations involved a Lockheed Constellation, the tri-tailed, four-engine transport model, where the flightcrew inadvertently flew into the water while executing a night visual approach to the airport on St. Thomas in the U. S. Virgin Islands. The accident involving Eastern's Lockheed L-1011 where the crew inadvertently let the airplane descend into Florida's Everglades Swamp while attempting to cope with an unsafe nose landing gear indicator light (the now famous, or maybe infamous, Flight 401), would never have occurred if the same set of circumstances had prevailed during daylight hours. In still another accident, authorities concluded the black hole phenomenon was involved in a fatal jet transport airplane crash at Pago Pago in the Pacific some years ago.

As more of these peripheral matters not directly concerned with the actual control of an airplane are discussed, it should become apparent that there is more to gaining flight experience than flying from point "A" to point "B" repeatedly, or making lots of takeoffs and landings. The really experienced pilot has accumulated many flight hours in a wide variety of airplane models under all manner of weather conditions. Additionally, he has knowledge about a host of other subjects or conditions that might come into play on particular occasions.

ISLAND HOPPING

A pilot was forced to ditch his airplane while he and his wife were island hopping in the Bahamas as part of a treasure-hunting game. The four-place airplane sank and was not recovered. The 33-year-old pilot, who sustained head injuries, expired before his spouse was found, and his body was not recovered. He had advised the nearest airport they were not going to make it after the engine failed. The U. S. Coast Guard rescued his spouse about three and one-half hours after the ditching. The survivor, a lovely, warm-hearted registered nurse, was interviewed ten days after the ditching while convalescing in a Miami, Florida hospital. Her mother was present when a tape recording of her account of the accident was made.

Under such circumstances, attempts were always made to be as gentle and understanding as possible. It was understood we would discontinue the interview any time the witnesses desired and they did not have to answer questions that gave them any concerns whatsoever. I also attempted to let them relate their experiences without interruption. These measures were utilized to avoid creating problems, emotional or otherwise, that might impede their recovery. However, on many occasions, it was obvious the witnesses were eager to tell their stories and were relieved to have it off their chests.

The paraphrased essence of the statement by the survivor was as follows: They flew from Florida to Grand Bahama Island where the airplane was refueled to capacity. The following day, they flew to Eleuthera Island where fuel was not available, a flight of one hour fourteen minutes duration. Two days later, they experienced engine failure about 15 minutes after takeoff while flying at 2000 feet below scattered clouds on a flight to Long Island, Bahamas. There was no fire, they had normal fuel and oil pressure readings, and there was fuel in both tanks when the engine malfunctioned. The pilot checked all the switches and valves and made repeated attempts to restart the engine without success. Upon descending through 500 feet, he advised the nearest facility they were not going to make it.

They landed tail first in three to four-foot seas, bounced a little, then came down hard in an upright attitude. The cabin door flew open on impact. After the survivor abandoned the wreckage, the pilot threw her the new life raft they had purchased for their island hopping adventure. The pilot then got out, but did not have time to retrieve the life vests or emergency locater transmitter (ELT) from the rear passenger seats before the airplane sank. The pilot was bleeding from a cut on the head, and his wife was unable to get the life raft inflated. The pilot kept holding onto her in his attempts to remain afloat. However, she was also experiencing much difficulty remaining afloat while continuing to try to inflate the life raft. Their struggles continued and she eventually became aware that her husband had expired. Realizing she would never be able to keep him afloat, she gave him a kiss and he drifted away after she released him.

Nearly exhausted, she continued to struggle with the liferaft. Some two, to two and one-half hours after the ditching, she finally got it partially inflated and it provided some buoyancy. About an hour later, she observed airplanes overhead and realized she had been spotted. However, she wondered why it took so long for the Coast Guard helicopter to arrive and lift her aboard.

While discussing specific points after her statement, she expressed concerns that the Safety Board would blame the engine failure on her husband. She was advised to have no worries whatsoever in that regard. The wreckage sank in ocean waters about 4,500 feet deep and would not be recovered. Additionally, the investigation was under the jurisdiction of the Government of the Bahamas and for such occurrences the Safety Board did not make a probable cause determination. She was also advised we would check with the Coast Guard to determine the factors relating to her rescue after she was spotted. Several days later, I called to tell her mother the U. S. Coast Guard's logs at the Miami, Florida Rescue Coordination Center showed she was spotted by a private airplane at 1:20 p.m.; a Coast Guard fixed wing airplane was overhead at 1:30 p.m.; and she was hoisted aboard their helicopter at 1:40 p.m.

She was still bedridden during the hospital interview and it was obvious the giving of such a vivid, step by step account of her ordeal had not been an easy task. Her poignant descriptions left lasting impressions on me. While reaching out to shake her hand before leaving, I told her she was a very brave young lady. After our handshake, she didn't release the pressure immediately. She clasped my hand only a few seconds, but the lingering pressure was perceptible. These are the occasions when I, and probably most other people, feel so helpless. We have an earnest desire to do anything possible to alleviate the situation. But then we have to face the reality that there is actually very little to be done other than express our sorrow and deepest sympathy. Many emotional situations were encountered when talking to, or conducting necessary business with, survivors and the next of kin of victims of aircraft accidents. However, the circumstances of this interview aroused a deep sense of compassion that remains most vividly etched in my memory.

While we were unable to determine the cause of the engine failure, there are lessons to be learned from this occurrence. Each and every occupant of light airplanes should wear his or her life vest at all times while on overwater flights. They are extremely hard to don in the confined cockpit and cabin spaces, and any ditching or other contact with the water surface will already have created enough anxiety and confusion. Additionally, when occupants fail to wear their life preservers, they may have trouble finding one before abandoning the aircraft. Another advantage is they're usually yellow, which greatly assists searching aircraft. Many folks may not realize people or objects on the water surface are extremely hard to spot, especially when the winds are churning up rough, whitecap-laden seas.

QUIET AND SERENE, A REALLY GREAT PLACE

I was dispatched to Nevis, a West Indies island, to assist aviation authorities for the Windward and Leeward Islands in their investigation of the crash of a U. S.-registered airplane. Nevis is a relatively small (50-square mile) island with a population of about 13,000. In accordance with International Civil Aviation Organization (ICAO) procedures, my title was Accredited Representative from the state of manufacture and registration.

I always attempted to render the degree of assistance requested by the foreign government's aviation authorities on such assignments. Normally, foreign investigations were more time-consuming for Safety Board investigators than similar occurrences in the United States. There were very few aviation accidents in the Windward and Leeward Islands' area of responsibility, and there was not a full-time investigative staff. For this occurrence, the Deputy Director for Civil Aviation from their headquarters on Antigua was the designated Investigator-in-Charge.

During my tenure in the Miami, Florida office, each investigator covered two or three accidents in the Virgin Islands every year. Having never visited the islands, my wife had a strong desire to accompany me on many occasions. But I always managed to find some reason why it would not be feasible. In addition to my Safety Board assignments, I had visited the islands several times while in the Marine Corps and, quite frankly, didn't find them to be all that exotic. Shopping on St. Thomas had become too expensive; whether well-founded or not, I was always apprehensive for my safety and well-being while on assignments in St. Croix; and considered Puerto Rico already overpopulated. However, in deference to the Chambers of Commerce on those islands, it is readily admitted I'm the world's worst sightseer. When I say the Grand Canyon and the Castles on the Rhine are about the only things I have seen that really left a lasting impression insofar as sightseeing activity is concerned, I'm not being facetious.

Travel to San Juan, Puerto Rico was on a commercial jet, followed by what we called a puddle hopper flight to St. Christopher, another British island about 100 miles east of St. Croix, where I met the Investigator-in-Charge. Nevis was about ten minutes away by air, but it was late afternoon and we remained overnight on St. Christopher. My first act after checking into the somewhat modest hotel was to call my wife and tell her to come down and meet us on Nevis. While I hadn't seen anything as awesome and magnificent as the Grand Canyon or the Castles on the Rhine, the friendly, straightforward manner of the citizens, who spoke with a distinct British accent, was most impressive.

The accident involved a light twin-engine airplane that crashed on takeoff after the pilot experienced a loss of power on the right engine. The operator, who was not a pilot, occupied the right cockpit seat. The front seat occupants were killed, but a passenger in the rear of the cabin survived. The commercial flight operated out of the airport on a fairly regular schedule. It was returning to the home base in San Juan, Puerto Rico with a cargo of fresh seafood and automotive components. Investigation showed the airplane was loaded 740 pounds over the authorized gross weight of 9,600 pounds. Witnesses heard the pilot complaining to the operator about the overloaded condition. However, the operator prevailed, and the pilot carried out his pretakeoff preparations.

The single asphalt runway at the Nevis Airport was only 2,000 feet long. Witnesses who had observed the airplane make previous takeoffs thought the airplane accelerated more slowly than usual. There were also expressions of anxiety and concern between some observers who heard the pilot's protests about the overweight condition. They noted that the individual blades of the right propeller were discernible during the latter stages of the takeoff roll. At the very end of the runway, they observed the airplane lift off abruptly and enter a shallow left bank. It impacted the ground on the left wingtip at a point 416 feet beyond the end of the runway. Thereafter, the airplane cartwheeled before coming to a stop.

Another witness who observed the airplane approaching head-on said the right propeller came to a complete stop before the initial impact with the terrain. Examination of the wreckage showed evidence that the right engine was not producing power, plus indications the pilot may have been shutting it down at the time of the crash. Several cylinders on the right engine showed evidence of improper combustion, but the precise causes of the complete engine failure were not determined.

Consideration of the total runway length, in conjunction with the computed speed to which the airplane could have accelerated, showed it was impossible to continue with a safe takeoff after the right engine failed. Since there was insufficient runway on which to stop, and the terrain beyond the end of the runway was unfavorable, an accident was inevitable. However, the consequences of the accident were made more severe by the pilot's failure to abort the takeoff.

My spouse thoroughly enjoyed our four-day stay on Nevis. Our oceanfront accommodations were within walking distance of the island's largest community. The low key activity was refreshing and she strolled about at her leisure. Hanging on the wall over my desk today are four sketches she painted, including one of the birthplace of Alexander Hamilton, our first Treasury Secretary. We also learned that Lord Nelson, the English Admiral, had a mistress on the island. Accordingly, he was a frequent visitor there when not chasing pirates about the Caribbean near the end of the 18th century.

Necessary transportation for the investigation was provided by the Investigator-in-Charge. He acquired use of a topless, four-wheel, low-slung, rover type vehicle without seat belts or a roll-bar. For me, it was a harrowing experience to speed about with him on the narrow, winding roads, and on several occasions he was asked to slow down so I could get my heart out of my throat. However, he was a most delightful and interesting host and my spouse and I thoroughly enjoyed our associations with him. One day while driving past a coconut grove in a low-lying seaside area, he remarked that the island's main town used to be located there. I asked: "What do you mean by 'used to be', what happened to it?" He explained that back in the 16th century, a tidal wave struck the island in the middle of the night and everything and everybody in its path were destroyed.

Upon completing our activities, we returned to Basseterre, St. Christopher, the capital city of the British self-governing territory of St. Christopher and Nevis. We had an audience with the Prime Minister, the Honorable Robert Bradshaw, during which we discussed the pertinent facts, circumstances, and conditions of the accident. He was very cordial and showed a keen interest in our investigative activities. Thereafter, we chatted a bit and he explained that they were interested in developing resort complexes. However, he stated that they wanted to move slowly so as to ensure the maximum benefits for their citizens. Continuing, he said: "We're in no hurry, we're eating pretty good now." Several years ago, I read that he died. Some resort development has been accomplished on the island, and my earnest hopes are that it has provided some meaningful benefits for their warm and friendly native inhabitants.

We remained overnight in Basseterre and meandered about the picturesque waterfront the following morning, which happened to be a Saturday. It was interesting to observe the curbside business dealings of the citizens who came out of the countryside to display the food and other articles they had for sale. We returned to San Juan, with a stop in St. Thomas that provided an opportunity to stroll about the shopping area in Charlotte Amalie, the capital. Thereafter, my spouse had an opportunity to sightsee in San Juan, Puerto Rico, while I completed the necessary investigative activities at the operator's home base. After returning home, she said she thoroughly enjoyed the trip and now understood why I had been reluctant to let her accompany me on trips to the U. S. Virgin Islands.

HAVE EXPERTISE, WILL TRAVEL

An accident in Georgia involved a light, twin-engine airplane owned by a large corporation engaged in Interstate Highway construction. It was climbing out in excellent weather after taking off from a nearby airport. Witnesses said it appeared one of the wings was folded upward when it dove into a lake, killing the pilot. After arriving, we went to the crash scene where only some floating debris had been recovered. Realizing the wreckage was secure on the bottom of the lake, we began collecting witness statements plus information concerning the pilot, airplane, and the purpose of the flight. In all my investigations, it was found beneficial to learn who was doing what, before examining the wreckage.

The principal officer for the highway construction company showed a keen interest in our investigation. After discussing the matter, he agreed to recover the wreckage. Impact forces resulted in significant breakup of the airplane, but three days later, the bulk of the wreckage was spread out in a secure area at the owner's facility.

The investigative team in this instance consisted of an FAA operations inspector and myself. When we failed to piece together the puzzle, I called my supervisor who had about 25 years' investigative experience. After discussing the evidence we had collected, he said: "Tom, I don't believe we have anyone in the office who can help you. You are free to call Washington, but I don't believe you will get any meaningful help from up there unless you can entice Jack Leak to come down."

The accident happened early in my career, but putting forth my boldest front, I called Mr. Leak and requested his assistance. Jack was a supervisor in our Washington, D.C. headquarters whom I had met while attending a basic investigative course. He said he couldn't come but would send another specialist from his division. I told him that wouldn't do, we had a very influential owner who had expended large sums to recover the wreckage, and we owed him our finest effort. We discussed the evidence, including the folded wing observed by the witnesses. He was also reminded he did not have to respond immediately; we would retain custody of the wreckage for the Safety Board until he could get away.

Remembering that my supervisor had said it would be necessary to have Mr. Leak come down, I pleaded with him to study the matter before making his final decision. Jack finally agreed, and about three hours later, he called back, asked for directions to the wreckage location, and said he would be down about eleven the next morning.

He arrived on schedule and was introduced to the owner and other team members. I jokingly told him he was now out in the field with the working folks and would have to shed his business suit. He said he brought coveralls but wanted to look around first. Less than 15 minutes after his arrival, he pointed to the left engine that had a short piece of the outboard wing spar attached. The fractured end of the spar piece was facing down about three inches off the ground, and the fracture surface was not visible. He said: "Tom, if you will have this engine turned over, we will find a fatigue crack on the fracture surface of the spar." When it was turned over we observed a classic example of a fatigue crack. That is where the smooth surfaced fatigue zone separates on a 90-degree plane until the remainder is incapable of carrying the load, Thereafter, the instantaneous fracture area is visible that breaks on a ragged, 45-degree angle.

The owner of the airplane was amazed. He remembered we had promised a detailed examination after the wreckage was recovered, but said he never would have believed we had anyone who could find the cause so quickly. After viewing the fracture, he said it was crystal clear, while remarking they broke many pieces of metal in their highway construction business that showed very similar features on the fracture surfaces.

Jack told me his studies showed previous problems with fatigue cracking at that particular location on the tubular wing spar. Accordingly, he figured he knew the problem area and had only to locate the correct piece.

After solving the puzzle for us, Mr. Leak took immediate actions that resulted in the FAA's grounding of the airplanes until remedial actions were taken. Jack's assistance probably would not have been required if the accident had occurred later in my career. However, I was never reluctant to ask for help. I always thought the quality of our investigative efforts was greatly enhanced by the use of knowledgeable specialists. Regretfully for the Safety Board, Mr. Leak departed from the Safety Board scene several years later after accepting a position with one of our military services.

While readying my investigative report, I noted an article in an aviation publication that was scornful of the FAA's grounding action. Investigators often encounter difficulty in efforts to ensure that news media personnel accurately report the factual information concerning an accident. Generally, difficulties were encountered when reporters not familiar with aviation tried to analyze or otherwise put in their own words the information we provided them.

To find such an inaccurate article in an aviation-oriented publication was appalling. It was so far off the mark that I thought it must have been written by a reporter who had previously been employed by one of those sensationalistic tabloids found in racks adjacent to the checkout counters in convenience stores.

At the conclusion of any inquiry, the Investigator-in-Charge is the best source for information concerning the facts, circumstances, and conditions of an accident. In accordance with our procedures, all the reporter had to do was ring me up to receive accurate information.

Reviewing the evidence again: We had witnesses who observed the airplane dive into the lake, with the portion of the left wing outboard of the engine extending upward. One of the witnesses was a pilot flying 1,500 feet above the terrain, who first spotted the airplane when it was about 1,000 feet above him in the dive. Examination of the wreckage showed a fatigue crack in the lower tubular spar just outboard of the left engine. The wing spar separated in an area where cracks had been found previously.

The FAA had issued an airworthiness directive (compliance is mandatory) requiring inspection of the fatigued area at 500 hour intervals. The latest inspection required by the airworthiness directive was accomplished six weeks prior to the accident. Records showed the airplane had only 23 hours in service since the inspection. Without question, there was every reason for the FAA to institute prompt and positive corrective measures after such a set of circumstances.

Let us now review several excerpts from the inaccurate article in the aviation publication:

> a. Pilot had reported at 4,000 feet only seconds before splash-down -
>
> b. FAA concluded the plane had broken up in midair and this brought on the grounding action -
>
> c. Later investigation disclosed the plane might have been buzzing boats at the time of the accident with the result that a wing had struck the water -

Subscribers deserved a more professional effort by the editors for the aviation publication. A copy of the article was forwarded to our Washington headquarters with the following comment:

> ***"This erroneous information illustrates the consequences of printing rumors when the true facts are unknown."***

NEWSWORTHY

A most newsworthy general aviation occurrence involved a helicopter crash in which the pilot and three news media personnel perished. On the morning of the accident, the flight departed from Miami, Florida on a flight to Cayo Lobos, an uninhabited island in the southern Bahamas, where a number of Haitian refugees were stranded. Two helicopters belonging to another operator also flew to Cayo Lobos the same morning with news media personnel on board. The situation on the island had become quite tense and newsworthy because the refugees had taken a stance whereby they were steadfastly refusing to be returned to Haiti.

According to occupants of the other helicopters, Bahamian authorities arrived at Cayo Lobos sometime after the helicopters. One of their first actions was to order the pilots of the helicopters to enplane their passengers and depart, leaving no doubts that they meant the helicopters should stay completely away from the vicinity of the island. The authorities were accompanied by armed security force personnel, and the helicopters departed immediately. The procedures the authorities employed to cope with the situation, and whether any persuasive or forceful measures were required, are unknown. However, the refugees were returned to Port-au-Prince.

The helicopters flew to the Congo Town Airport on Andros Island where prior arrangements had been made to obtain fuel. The company's operating procedures for the pilots of the other helicopters did not permit them to conduct night overwater flights. Accordingly, they took on only enough fuel to fly to Nassau, in order to leave a greater quantity for the pilot involved in the accident, who intended to return to Miami, Florida.

All three helicopters departed shortly before dark. The pilots of the helicopters bound for Nassau reported they encountered overcast conditions with intermittent rain showers, the same weather that had persisted throughout the day. Satellite photographs for the time of the flight showed overcast conditions at Andros Island, with the cloud formations extending over portions of the direct route to Miami. There was no record of communications with the flight bound for Miami after the departure. The flight failed to arrive and the Coast Guard conducted an extensive search, with negative results.

Seventeen days after the accident, a piece from an airplane was found in the surf at the west end of Grand Bahama Island. After being flown to Miami, it was positively identified as wreckage from the missing helicopter. The ragged, five-foot square piece of the fuselage structure from directly beneath the cockpit seats showed all or part of the last four registration numbers of the missing helicopter. Neither the occupants nor any other components or debris were ever found. The recovered piece of wreckage showed extreme impact forces had resulted in an extensive breakup of the helicopter. Such a degree of breakup is consistent with impact during uncontrolled flight. However, since the remainder of the wreckage was not recovered, it was impossible to determine whether a failure or malfunction of some component or system had been involved in the accident.

News media personnel always appeared to be extremely dedicated to their profession. They are constantly striving to get from here to there in the shortest period of time in order to meet some deadline. To achieve their goals, they oftentimes exhibit a nearly blind faith and trust towards those able and willing to provide some necessary service. There is little doubt the reporters who lost their lives in this tragic accident did not have full knowledge of the shortcomings in the pilot's operating certificate, his individual pilot qualifications, the helicopter equipment deficiencies, and the pitfalls that might be associated with the conduct of such a night flight in the prevailing weather. In the event they were fully apprised of those circumstances, they too would probably have elected to stay overnight in Nassau before returning to Miami.

Investigation showed the pilot was the sole owner of the company operating the helicopter. The helicopter involved in the accident was his only aircraft. Bahamian regulations required that night flights be conducted under instrument flight rules. The pilot did not file a flight plan, was not instrument rated, the helicopter was not equipped for instrument flight, and the pilot's application to the FAA for the authority to conduct such commercial operations was still pending. In contrast, the pilots of the other helicopters were instrument rated, but their aircraft were not equipped for instrument flight, and they proceeded to Nassau in order to comply with their company's procedures and Bahamian regulations.

A review of the evidence shows the pilot was on a perilous journey. Once night settled in, he would have experienced difficulty conducting the flight via visual references even when clear of clouds. Under the overcast sky with no lights on the water, he would not have seen the horizon and the ocean surface would most likely have been completely obscured in the darkness. Noninstrument-rated pilots are subject to becoming spatially disorientated, meaning confused as to which end is up, under such conditions. Thereafter, a complete loss of control can, and often does, occur. The Safety Board's files contain numerous occurrences where overzealous, unsuspecting, or inexperienced pilots have terminated their flights catastrophically under similar conditions or like circumstances.

- -

THE FLOWER OF OUR YOUTH

Violence, death, destruction, sorrow, grief, etc., these are the ingredients that brew up the most newsworthy occurrences, especially on the local scene. In case you have any doubts about this assertion, just watch your local news tonight, or any night for that matter. Around my house, it's sometimes referred to as the half-hour disaster report. Accordingly, when an accident results in multiple fatalities to what might appropriately be called "The Flower of our Youth," it garners detailed and widespread coverage. Such was the case for an accident involving an airplane belonging to the Congressional Flying Club that crashed into the side of a mountain near Anniston, Alabama, killing four aides to Alabama congressmen.

With five passengers on board, a non-instrument-rated congressional aide had flown the airplane from Washington to Alabama for the Thanksgiving holidays. Upon learning the return flight could not be conducted under visual flight rules, he acquired the services of an instrument-rated pilot to accompany himself and four of the passengers back to Washington in the six-place, single-engine airplane. The return flight originated in Monroeville, Alabama, and made a stop in Birmingham to enplane passengers. Thereafter, it departed for Anniston where the remaining passenger was to board the flight before continuing on to Washington, D. C.

The instrument-rated pilot-in-command who had been hired for the return flight let the noninstrument-rated congressional aide fly the airplane from the left seat. Such a procedure is often utilized, especially in the training environment. While it violated no FAA regulations, I always had reservations about such a practice when passengers were on board. The airplane was equipped with dual controls, but flight instruments were installed only on the left side of the instrument panel.

The 20-year-old instrument-rated pilot-in-command in the right seat reported 520 total flight hours, including 35 hours' actual instrument flight experience. The 19-year-old congressional aide flying the airplane from the left seat had about 200 total flight hours. From a flight experience standpoint, these are not impressive totals.

The weather at Anniston showed a 1,000-foot ceiling with seven miles' visibility. After passing Talladega, Alabama, the Birmingham controller cleared the flight for the localizer instrument approach to Anniston. Anniston did not have an airport control tower or any other air traffic control facilities. The Birmingham controller was unable to provide radar monitoring of the localizer instrument approach because of terrain features in the vicinity of Anniston.

A localizer instrument approach provides directional guidance only to the flightcrew. Precise descent guidance is not available when conducting a localizer approach. Proper execution of the localizer instrument approach procedure required that the flight turn right, a direction taking it away from the airport, after initially intercepting the localizer course.

The instrument-rated pilot-in-command, the only survivor, said they turned left towards the airport upon intercepting the localizer course. Thereafter, they descended from their assigned altitude of 4,000 feet to 1,600 feet. In response to my question as to where he thought the airport was at the time of the crash, he replied he thought it was ahead of them. For all intents and purposes, the investigation was over insofar as determining the cause of the accident was concerned. The flight was almost over the airport when they intercepted the localizer course. Thereafter, they overflew the airport that was not visible because of the low ceiling, and descended into rising terrain about six miles to the northeast.

The statement by the survivor greatly influenced the scope and magnitude of the investigative effort. Making a blunt analysis, we concluded that the pilots couldn't have fouled up the instrument approach procedure any better. But don't leave with the impression that such mistakes are reserved solely for the inexperienced aviator. The TWA Boeing 727 crash in Virginia that occurred while the pilots were making an instrument approach to the Dulles International Airport outside Washington, D.C. happened when the flightcrew made a premature descent into the backside of a mountain between them and the airport. Additionally, there have been many other disastrous accidents where experienced pilots have failed to properly execute precision instrument approaches to airports.

Only a year before this accident, I had investigated another fatal occurrence right at Anniston, involving a well qualified, instrument-rated pilot who flew his airplane into the same mountain while executing the localizer instrument approach procedure to the Anniston airport. There is really no mystery about flying a precision instrument approach to any airport. Pilots are required to possess the instrument approach procedure chart that provides both written instructions and a diagram, showing the exact manner in which the approach is to be executed.

Upon completing the investigation, I made a recommendation proposal to the Safety Board that called for a revision to the FAA regulations. It would have required that airplanes operating in instrument weather with passengers on board be flown by an instrument-rated pilot with a full set of flight instruments on his side. However, my proposal wound up in that old wastepaper basket file in Washington. The reviewer contended it would not be possible to enforce such a regulation.

Enforcement was not the issue. History will show somewhere in excess of 95 percent of our citizens are going to obey our basic laws and regulations. To this day, I believe my recommendation was warranted. If the provisions of my proposal had been in effect at the time of the flight, the instrument-rated pilot-in-command would have been required to fly the airplane from the left seat. While that may not have prevented the accident, it would have been a better arrangement than that which existed at the time of the crash.

All government agencies obtain their funding from Congress. Accordingly, it is not surprising that those agencies give immediate attention, and put forth a maximum effort, when responding to congressional inquiries. At the Safety Board, much of the burden for such a maximum effort is borne by the investigator. Needless to say, there was a near flood of congressional inquiries immediately after this occurrence and I spent an inordinate amount of time in Anniston just being "on the scene."

The Safety Board normally encourages its investigators to respond to news media inquiries by giving factual information, without speculating as to the probable cause, of course. However, I was effectively muzzled in this instance. The only response I was suspposed to give was, "The investigation is ongoing," while referring callers to our Public Information Office in Washington. If you have any concerns about a matter under investigation, and do not appreciate the manner in which the Safety Board is responding, write your Congressman or Senator. The Safety Board may not end up doing what you desire, but you can rest assured they will give your representative a comprehensive response. And such a procedure also works well with most other governmental agencies.

- -

FAR-FETCHED

I was dispatched to the Turks and Caicos Islands to assist in the investigation of an accident involving a U.S.-registered, four-engine transport airplane. This island nation, located between the Bahamas and Puerto Rico, is not a very popular vacation retreat and only thrice-weekly flights were available between it and Miami, Florida. When our office initially received notification of the accident, we were informed the airplane was transporting a cargo of fresh meat and seafood from the Central American nation of Panama. However, after arriving on scene, we discovered the flight was also carrying a significant quantity of marijuana.

The Turks and Caicos Islands have retained some ties to Britain, and I had an audience with the Governor who is appointed by the Crown. He had a rather spacious office, somewhat detached from the local community, where life-sized pictures of Queen Elizabeth and Prince Philip were on display. The measures Britain uses to locate and train its foreign service representatives apparently produces highly qualified candidates. I was impressed by their dignity and the aura of pomp and circumstance that seemed to permeate their surroundings. While exhibiting a stately manner, the Governor was warmly cordial and we had a very pleasant conversation. He mentioned that it appeared we had a drug-related occurrence. He was advised I probably wouldn't have left Miami if aware we were dealing with a pot hauler. However, since I was already there, we would render whatever assistance their Investigator-in-Charge desired.

The airplane was destroyed when it crashed on South Caicos Island. Two flight crewmembers and two extra crewmembers, the only persons on board, escaped without serious injury. A sophisticated twin-engine Lear jet airplane came down from the states the following day and whisked the airplane occupants away before the marijuana was discovered.

I always had difficulty garnering much enthusiasm for the investigation of drug-related occurrences. My basic approach was to expend every effort to render assistance and cooperation to law enforcement personnel and other responsible agencies. Then, after collecting the necessary statistical data, the wreckage examination usually involved taking some photographs of the scene and standing around kicking the debris, while uttering a prayer of thanks if those involved in the operation had cashed in their chips and gone on to their final rewards.

The home base of the airplane was Panama and the operator was reported to be an influential Panamanian. The application to our FAA for a certificate to operate the airplane in foreign air commerce had been approved. The application showed they were purveyors of fresh meat and seafood products in Central and South America and the Caribbean area. They were not authorized to operate the airplane in the United States.

It is probably difficult to understand how such an airplane could ever be registered in the United States. Possibly the best means of explaining it is to draw upon an analogy whereby all the cruise ships operating out of Florida are registered in Liberia or Panama or some other foreign country. Obviously, the owners register their ships in foreign countries to avoid being subject to our stricter maritime laws. Well, the same thing is true with airplanes, only in the reverse. Now we understand how it could carry a United States registration, but please don't ask why we permit it, because I am unable to explain the logic behind such a practice.

The pilots said they experienced operational difficulties with the number four engine shortly after takeoff.

Approaching Hispaniola, they experienced electrical problems and after cancelling their instrument flight plan, turned off all electrical equipment. In the prevailing visual weather, they overflew the international airports at Santo Domingo and Puerto Plata on Hispaniola, and continued on to Grand Turk Island.

The Grand Turk Airport Control Tower was in service but the controller was unable to establish communications with the flight when he observed it overfly the airport. The time was near dusk, and the controller took measures to have the runway lights illuminated at the South Caicos Airport.

During a post-crash interrogation, the flightcrew stated they were unable to see the runway lights upon arrival over the island. The tower controller at South Caicos spotted the unlighted airplane overhead and observed it flying off in a southwesterly direction. The captain said he elected to lighten the load to extend the range. so they tossed some of the cargo into the ocean. Interestingly, that was where the 35 bales of marijuana were found. Returning to South Caicos, they experienced engine failures due to fuel exhaustion and crash landed the airplane about two miles south of the airport.

The pilots were U. S. citizens and they completed the Pilot/Operator Aircraft Accident Report required by the Safety Board. They also related rather detailed accounts of all aspects of the flight and the night crash landing on the island. Obviously, they omitted any reference to, and denied any connection whatsoever with, the 35 bales of marijuana found in the ocean southwest of South Caicos Island the day after the accident.

Their statements were so far-fetched and full of incongruities that difficulty was experienced discerning whether I was attending a liars' convention or visiting Fantasyland. However, I was aware they would be subjected to no adverse actions. From experience, it had been learned law enforcement personnel's hands are pretty well tied unless they apprehend suspects with the contraband cargo on their airplane, or otherwise physically in their possession.

One of our drug enforcement agents telephoned me after I returned to Miami. He requested information concerning the accident, saying they had a dossier on the operator, who reportedly had a history of dealings in drugs. The agent's office was just down the street, and we delivered a copy of everything in our file. After a brief review, he said that was just what they needed. He was asked to give us a call if we could be of further assistance. A couple of months later, I called to see what progress had been made. The agent replied he wasn't able to conduct follow-up actions because their office had a maximum effort on another case that required all of his time and attention.

- -

NOT QUITE BY THE BOOK

The helicopter crashed into the Potomac River just south of the 14th Street Bridge in Washington, D.C. The site was opposite the National Airport where the helicopter had departed with three passengers aboard on a commercial aerial survey and photographic mission of property along the riverfront. The weather was beautiful and numerous witnesses observed the helicopter hovering 150 to 200 feet above the water surface. The right cabin door had been removed and some witnesses observed that a passenger taking pictures had his feet hanging outside. The engine noise suddenly ceased, and the helicopter commenced rotating to the right. It maintained somewhat of a level attitude while making a near vertical descent into the water. The severe splashdown forces killed the three passengers and seriously injured the pilot.

Depending upon the make, model, and configuration, every aircraft has its operating parameters and limitations. Helicopter handbooks have a diagram showing an area from about 10 to 400 feet above the terrain or water where hovering flight is to be avoided. The underlying reason behind the diagram, which every helicopter pilot is expected to know and understand, is that an airspeed of about 40 knots is required at those heights if the pilot is to maintain sufficient rotor revolutions to effect a safe landing in the event of a complete loss of power. Basically, the airflow through the rotor system created by the airspeed produces a windmill effect that keeps the rotor turning. The pilot must take other measures to cope with a sudden and complete loss of engine power; however, those are accomplished almost as a reflex action by the experienced helicopter pilot.

In the event of a sudden engine failure while the helicopter is hovering, a rapid decrease in rotor revolutions occurs because of insufficient airflow through the rotor system. This was the exact happening in the accident under discussion and accounts for the rapid, vertical descent of the helicopter into the Potomac River. When hovering at heights of 400 or more feet above the terrain, the pilot has enough altitude in which to dive the helicopter to gain airspeed. This will result in an increase in rotor revolutions and the pilot is then able to effect a safe landing.

This precept is highlighted in the U. S. Navy's Ten Commandments of Helicopter Flying that states: "Thou shalt maintain thy speed between ten and four hundred feet, lest the earth rise and smite thee." The moral being: helicopters are not to be utilized routinely as elevators.

In helicopter terminology, upon experiencing a complete loss of power, the pilot goes into an autorotation whereby he trades altitude for enough airspeed to maintain the desired rotor revolutions. The spinning rotor has enough inertia for a one time, soft landing touchdown. After a flare is executed to stop the forward momentum, the pilot increases the pitch on the blades to use the rotor's inertia to cushion the landing. This causes a rapid drop in rotor revolutions, but hopefully the helicopter is safely down by that time.

For clarity, it should be understood that when flying directly into a headwind, an aircraft does not distinguish between airspeed and wind speed. Therefore, hovering a helicopter into a 40-knot headwind is equivalent to maintaining a like airspeed under calm wind conditions. For the accident under discussion, very light winds were prevailing. Examination of the wreckage showed the pilot lost all power to the rotor because of the failure of a gear component in the engine. Since the helicopter was hovering without any appreciable airspeed, a rapid decrease in rotor revolutions occurred which the pilot could not regain because of the low height above the water surface.

Passengers contracting for air-taxi, photographic, or local sightseeing flights are buying the services of the pilot and the wear and tear on the machine. Normally, passengers are not well versed in aviation matters nor are they aware that the FAA regulations for photographic and local sightseeing flights are not nearly so stringent as those for air-taxi and other types of commercial operations. Most airplane passengers assume they will be flying in an airworthy aircraft with a pilot who will exercise prudence and good judgment to ensure their complete safety. Painfully, that is not the way it is on many occasions.

The assigned pilot may be prone to take some chances to show off his flying skills or for other reasons. Additionally, passengers may be dealing with shoestring operators, not far removed from financial collapse, who are operating aircraft that are not in compliance with mandatory maintenance requirements, and whose liability insurance coverage may have been cancelled a few days or weeks earlier. It is also important to remember material failures leading to catastrophic consequences can occur on even the best maintained aircraft.

Regretfully, comprehensive measures for dealing with such potential pitfalls are not easily defined. You should know your operator whenever possible. When arranging for such flights, never be reluctant to tell the pilot you are expecting him or her to adhere to safe and sound operating procedures while never compromising safety for any reasons whatsoever. In the event you must utilize small operators with whom you are not familiar, it is a good practice to carry enough personal insurance to protect your next-of-kin in the event of unforeseen eventualities. The first thing most liability lawyers accomplish is a study to determine whether there is a deep money pocket they can get their hands into in the event they win their case. If there's no deep pocket, most won't pursue the matters on a contingency basis.

WHERE EXPERIENCE DOESN'T COUNT

This was an accident where we came within 250 feet of making a significant contribution to aviation safety. That is the distance the airplane missed one of the satellite terminal buildings occupied by a large number of passengers when it crashed on the Tampa, Florida international airport. The pilot was attempting to take off in weather where both the ceiling and visibility were zero.

The commercial airlines were effectively grounded because the weather was below their prescribed takeoff minimums. However, FAA regulations do not specify any takeoff minimums for noncommercial, general aviation flights. In those instances, the decision as to whether or not a flight will depart is left to the discretion of the pilot-in-command. Accordingly, the regulations let less qualified pilots go aviating when the true professionals are grounded.

After the investigation, a recommendation proposal calling for the FAA to establish takeoff minimums of not less than 200 feet ceiling and one-half mile visibility for such flights was forwarded to our Washington headquarters. However, my proposal found its way into the old circular wastepaper file. We wrote it off as just another instance of no tombstones, no corrective measures. Now, if the airplane had continued another 250 feet and slammed into the terminal building with catastrophic consequences, the powers that be in Washington, D. C. would have been scurrying about making revisions to the regulations without any help from us.

It was almost by the grace of almighty God the flight did not end with more catastropic consequences. The four occupants were injured and several automobiles and some fencing were damaged, but there were no injuries to other persons and no additional property damage. The occupants were returning home on a pleasure flight. The pilot had filed an instrument flight plan to the initial en route stop in Montgomery, Alabama. The wreckage came to rest on an embankment just short of the satellite terminal ramp. An unoccupied McDonnell-Douglas DC-9 passenger airplane was parked on the ramp 20 feet in front of the wreckage.

The 54-year-old pilot, a fine gentleman and chief executive officer of his company, rendered the utmost cooperation during the course of the investigation. His straightforward attitude and comprehensive statement focused in on the factors related to the accident and materially narrowed the scope of the necessary investigative effort. The substance of the pilot's statement follows: "Immediately after liftoff, the airplane settled a little when he retracted the landing gear and wing flaps. He thought he may have aggravated the situation when he engaged the automatic pilot before reestablishing a climb. Although the takeoff was made on runway 36, the pilot incorrectly set the automatic pilot heading indicator to 330 degrees. When he engaged the automatic pilot in the heading mode immediately after takeoff, the airplane began turning left towards the 330 degree heading. Thereafter, the airplane settled back into the terrain after turning away from the runway heading."

In this instance, the pilot did what many other pilots do. Believing the automatic pilot will fly the airplane better than they can, they make maximum use of it upon encountering actual instrument weather conditions. Although not a recommended procedure during the initial climbout after take off, in those instances where the automatic pilot is used correctly, the practice usually turns out okay. However, it can be fraught with disaster, especially when embedded thunderstorms or other weather phenomea are encountered that result in severe turbulence. The turbulence can cause the airplane to be tossed about into an attitude where the pilot must take over from the automatic pilot to effect a recovery. And it should not be necessary to further speculate how this might be injurious to the occupants' lifestyle, especially when the pilot is not proficient in instrument flying procedures and techniques.

WITNESS FOR THE DEFENSE

Federal regulations preclude the use or admission into evidence of Board accident reports in any suit or action for damages arising from an occurrence. Additionally, Safety Board employees are not permitted to appear and testify in court in actions or suits for damages resulting from accidents. Their testimony may be solicited through depositions or written interrogatories after the litigants obtain approval from the Board's General Counsel. Normally, depositions are taken and interrogatories answered at the Board's office to which the employee is assigned. However, Safety Board employees are permitted to appear and give testimony in criminal proceedings on some occasions. Such was the case when our General Counsel called to inform me he had approved a request by the Justice Department for my appearance in a criminal case to be tried in a relatively small city in one of our midwestern states.

The federal case concerned, among other issues, an accident involving a light, twin-engine airplane I had investigated. Our General Counsel reminded me of the limited scope of my authorized testimony as stipulated in the Safety Board's regulations. Basically, that meant my testimony was limited to the factual information obtained during the course of the investigation, including factual evaluations embodied in my investigative report. I was not permitted to give testimony beyond the scope of my investigation; to give expert opinion testimony; or to testify concerning the Safety Board's report containing its probable cause determination. He stated his office would forward the original of my investigator's factual report for reference purposes. He suggested I read into the record the statement we routinely used before testifying in deposition proceedings to ensure that the court was fully informed as to the scope of my permissible testimony.

For clarification, there are two reports for each accident. The investigator's factual report with its attachment, contains the evidence collected during the investigation. The Safety Board's report highlights the factors involved in the accident, and provides its probable cause determination.

Necessary liaison was conducted with the assistant U. S. attorneys handling the case to determine an appropriate arrival date and to coordinate other matters. The attorneys were satisfied that my factual report was being brought for reference. However, I expended every effort in this arena to avoid possible embarrassment while wrangling with those peripheral questions that are always asked. We've all probably heard the somewhat facetious remark that an excellent method of hiding our ignorance is to stop wagging our tongues. Obviously, that's not possible when testifying, and the court reporter transcribes every word and phrase exactly as it's stated. Based on past experiences, dating back to the review of the transcription of my very first deposition, I was adhering to a firm commitment made to myself to never again come across sounding so ill-prepared. Accordingly, I boned up on every facet of the investigation, conducted much study of background material, and arrived with far more reference material than would ever be required to testify only as to the factual evidence contained in my investigative report.

After I checked into the motel, the attorneys came over for a get-together. It was a pretty involved case concerning, among other matters, some drug-related charges. When asked how the airplane accident was involved, they replied that one of the charges was that the defendant had sabotaged the airplane. I told the attorneys, "That is preposterous, and I can give no testimony whatsoever to support such a charge."

They replied, "Oh, we know that; you are being called as a witness for the defense." Being somewhat perplexed, I asked for an explanation. In essence, they went on to say they had about 40 charges against the defendant and didn't want to agree to drop any before the trial. They remarked that judges usually toss out a few of the alleged offenses, and they figured the sabotaging charge would be a likely candidate among those to be dismissed.

Being so much better educated into the ways and means of our court system, the attorneys were asked what testimony they intended to elicit. They said only the factual evidence contained in my report. When asked who would be on the jury, they replied local residents and farmers, and their wives and daughters. The occurrence involved a Ted Smith model 600 Aerostar airplane that, in my opinion, has one of the most complicated of all light airplane fuel systems. I informed them that we couldn't be discussing a more difficult occurrence because it involved a double engine failure related to fuel system mismanagement by the pilots. The attorneys were further advised that anyone in the courtroom who didn't fully understand the fuel system would not be able to understand or comprehend what we were talking about.

This was not meant to imply that there was anything defective about the fuel system. It was explained that no fuel system is all that complicated once a person understands it. After conferring on the matter, it was agreed I would provide the necessary instruction for the court.

Thereafter, it was learned that no one involved with the case had a significant aviation background. The only persons in the courtroom with extensive knowledge and experience in small airplane operations would probably be the defendant and myself. So another judgment call was made. Sometimes it appeared I spent 20 years making such decisions that oftentimes infringed upon the

Board's regulations. However, if ever challenged, it was considered I could lean on the reference to permissible testimony by its investigators in the Board's regulations that stated: "...including factual evaluations embodied in their factual reports."

Anyway, the attorneys were told they could ask me anything they desired if such an arrangement was agreeable with both sides. I would give expert opinion testimony and explain anything known about aviation matters in general, or the airplane involved in the accident, in an effort to be as helpful as possible. It was rationalized that I had come too far and made too much preparation to refuse to use my knowledge and experience to enlighten the court while sitting there responding: "I'm not permitted to answer that question." However, it was made acid clear that no testimony would be given regarding the cause of the accident, nor would we discuss the Safety Board's report containing its probable cause. The Safety Board would consider any such testimony by its investigators to be an intrusion into its restricted and sacred domain. And rightly so, I might add.

The attorneys were advised we would need a schematic drawing of the fuel system as a training aid. A blackboard was available, and it was decided they would call me before lunch, get the preliminaries out of the way, and then ask me to draw the schematic during the noon break.

If we ever have absolutely foolproof aircraft fuel systems, meaning ones like those in our automobiles where there are no valves and any fuel remaining would be available to the engines, there would be a significant improvement in the aviation accident rate. Putting it another way, you'd be an odds-on favorite to win a bet that every Investigator-in-Charge with at least a year's experience with the Safety Board has covered an occurrence resulting from fuel system mismanagement by the crew.

Such accidents occur more frequently than television ministers can dream up another contribution gimmick. Although having a foolproof aviation fuel system would be akin to living in that Utopian paradise somewhere beyond the end of the rainbow, the design of such a system is probably impossible.

Even the simplest fuel system on a single-engine airplane with only one tank has a manual ON/OFF valve. When two tanks are installed, the fuel selector positions are normally OFF, Left Tank, or Right Tank; however, some of the smaller models have incorporated a BOTH position which puts us about as close to that never/never land as we'll ever get. Thereafter, the complexity of fuel systems generally varies proportionally as more engines or fuel tanks are installed. For the light twin-engine airplane environment, we can move to the top of the class insofar as fuel system complexity is concerned by studying the model 600 Aerostar involved in the subject accident.

The airplane has three fuel tanks, one in the fuselage and one in each wing. Fuel from each of the tanks initially flows into its own sump. These sumps are actually small holding tanks located in the fuselage center section below the fuselage tank. One-way check valves are used in the sump assemblies to prevent fuel from feeding from one tank into another. Four electrically operated fuel shutoff valves are mounted on the sump assemblies. The valves are controlled by fuel selector switches in the cockpit for each engine. The selector switch positions are OFF, ON, and X-FEED (CROSSFEED). When in the ON position, fuel is supplied to the engines from the respective wing tank (i.e. right tank to right engine) and the fuselage tank simultaneously. The system is designed to provide about 16 gallons of fuel remaining in the fuselage tank when the wing tanks run dry. For this investigation, it's important to note that 16 gallons of fuel were drained from the fuselage tank during the examination of the wreckage.

The X-FEED position permits fuel from the opposite wing tank to be supplied to an engine (i.e. left wing tank to right engine.) When the X-FEED position is selected, the fuel in the fuselage tank is not available to the engine. (Remember, we found 16 gallons of fuel in the fuselage tank.) The X-FEED position is designed for use in emergency situations such as an engine failure, or in other unusual circumstances where there might be an uneven supply of fuel in the two wing tanks. In the event the X-FEED position is selected for both engines, **the fuel in the fuselage tank will not be available to either engine.**

A three-position selector switch is located adjacent to the fuel quantity indicator gauge on the copilot's instrument panel. The switch allows the crew to select either the left or right wing tank and read the quantity of fuel remaining. Upon releasing the switch, it is spring loaded back to the TOTAL position that shows the fuel remaining in all three tanks.

The above information does not encompass everything the flightcrew needs to know and understand about the fuel system that incorporates an engine-driven fuel pump and electric boost pump for each engine; a fuel vent and expansion system; strainers; filters; drain valves; operating warnings and restrictions; etc. However, this brief description should provide enough background for the reader to understand and analyze the factual evidence found during the investigation.

The flight departed from Bridgeport, Connecticut, and crashed near Jacksonville, Florida five hours thirty three minutes later. Good visual weather conditions prevailed and the pilots did not file a flight plan. No record was found that they communicated with any facility prior to their arrival in the Jacksonville area shortly after midnight. The airline distance is about 800 miles and it was speculated they made an en route stop because the elapsed time was greater than required for the flight and also sufficient to exhaust the fuel supply. Efforts to retrace the flight path in such instances are akin to searching for the proverbial needle in a haystack because the pilots could have selected any of several routes and landed at hundreds of en route airports. Several months after the accident, it was learned that the airplane had landed in Savannah, Georgia. However, the exact arrival and departure times were no longer available because, unless warranted by an accident or other abnormal situation, FAA records are destroyed after 30 days. A check of refueling records failed to show that the aircraft was serviced while in Savannah.

The pilots called Jacksonville Approach Control when 30 miles north of the airport. The controller established radar contact and the flight was subsequently cleared to land on runway seven. Shortly after acknowledging that they had the field in sight and being cleared to land, the crew reported, "we've got both engines out." There were no further communications with the flight. The wreckage was found the following morning in a densely wooded area about one and three-quarters miles west of the airport. The airplane was destroyed and the two pilots were killed.

Both wing fuel tanks were ruptured during the crash sequence. Physical examination of the positions of the electrically operated fuel selector valves showed both engines to be in the X-FEED positions. Sixteen gallons of fuel were drained from the intact fuselage tank. Remember, fuel in the fuselage tanks is not available to either engine when both selectors are in the X-FEED positions, and the system is designed to provide about 16 gallons of fuel remaining in the fuselage tank when the wing tanks run dry. Examination of the wreckage and operation of both engines on a test stand showed nothing that would have adversely affected the operation of either engine if supplied with an adequate fuel supply.

The flight logbook for the airline transport-rated pilot in the left seat showed in excess of 10,000 flight hours, but only one prior flight in the model 600 Aerostar. However, his spouse reported that he had flown the airplane on several other occasions. The logbook for the single and multi-engine-rated private pilot in the right seat was not received. Fifteen-months-old FAA records showed he had accumulated 150 total flight hours. FAA regulations permitted the airplane to be flown with only one pilot on board. It was not determined which pilot was flying the airplane, or acting as the pilot-in-command, at the time of the accident.

Fuel system limitations contained in the FAA Approved Flight Manual for the airplane were as follows:

1. Each operating engine fuel selector must be in the ON position for takeoff, climb, descent, approach, and landing.

2. The fuel selector X-FEED position is designed only for lateral trim (those occasions where there may be unequal fuel quantities in the wing tanks for some reason), and single-engine emergency when it is necessary to utilize fuel from the opposite wing tank. Use of the fuel selector X-FEED position is restricted to level coordinated flight only.

3. Do not double X-FEED. (Which was occurring in this instance).

Records showed the airplane had not been maintained and inspected in accordance with FAA requirements. The required annual inspection had not been signed off by an authorized inspector, and necessary alterations or revisions contained in a mandatory airworthiness directive issued by the FAA had not been complied with. Compliance with the airworthiness directive would have required the installation of the following placards in the cockpit:

X-FEED LEVEL FLIGHT ONLY

DO NOT DOUBLE X-FEED

A discussion of the role of the Authorized Inspector (referred to as an "AI" in the trade) in aircraft maintenance programs might be beneficial. Almost everyone has taken an automobile to a garage for some specific problem and, after writing a sizable check for the replacement of expensive parts, had the identical problem manifest itself before arriving back home. It's problematical whether we received any monetary adjustments, and about the only damage the garage owner can sustain might be a blot on his business reputation. Usually, the mechanic who performed the work is not required to possess any certificates or licenses.

That's not the case in aviation. A certificated aircraft and powerplant mechanic may sign the logbooks for specific repairs he makes; however, sign-off for the required annual airworthiness inspection can only be accomplished by the holder of an Authorized Inspector certificate. It might be said an Authorized Inspector's certificate signifies that an individual has demonstrated he is qualified to move to a higher rung on the maintenance ladder. When an Authorized Inspector signs off for an annual inspection in the aircraft logbooks, he is certifying that the airplane and engines are airworthy, and all required airworthiness directives, modifications, revisions, and installations have been complied with. It is not unusual for the mechanic or inspector who last signed the aircraft logbooks to be named a defendant in liability suits arising out of accidents. The above makes it difficult to understand why aircraft mechanics sometimes receive less compensation than those who work on our automobiles.

It is the owner's responsibility to assure that an airplane is airworthy and the required logbook entries are made before it is flown. Normally, the owner is relieved of such responsibility if he delivers the aircraft logbooks to the maintenance facility, and it is not unusual for the logs and records to remain in the custody of the facility that maintains an airplane.

Back to our accident. In the event the required annual inspection had been conducted, the authorized inspector would not have signed the records until assured the mandatory airworthiness directives had been complied with. That would mean the fuel warning placards required by the airworthiness directive would have been installed on the instrument panel. Be that as it may, nothing relieves the pilot-in-command from the requirement of possessing in-depth knowledge of the airplane, including the fuel system and the fuel system limitations contained in the FAA-approved flight manual. Accordingly, the investigation showed deficiencies on the part of the owner for not having the required inspection performed, and of the pilot-in-command for his inadequate knowledge of the airplane's fuel system.

My testimony at the trial went rather smoothly and there were no adverse proceedings during the three hours I was on the witness stand. Everyone in the courtroom showed intense interest as the comprehensive explanation of the fuel system was presented, followed by a thorough discussion of the relevant facts, circumstances, and conditions of the accident. This was one of those occurrences where anyone, regardless of any prior aviation experience, has to come to the conclusion that the engine failures were related to fuel system mismanagement by the crew. Accordingly, it wasn't necessary to discuss in the courtroom the cause of the accident. Reflecting on my testimony, it was apparent that the decision to answer almost all questions had expedited the proceedings and assisted the lawyers on both sides in their efforts to provide the necessary background information for the jury, albeit some of the questions answered went beyond the scope of the Safety Board's permissible testimony.

Probably like many others, I always feel some anxiety, and am a bit apprehensive, when personally involved in legal proceedings. At the conclusion of my testimony, the court took a brief recess. While I was unwinding in the corridor, the presiding Federal Judge introduced himself and congratulated me for my testimony. Essentially, he said I had conducted a thorough briefing for the court on a rather difficult subject and everyone in the courtroom fully understood the matters under discussion. He went on to say there were occasions where he had difficulty following the testimony of some expert witnesses. I expressed my sincere appreciation for his kind remarks and we engaged in some small talk before parting.

While the above account completes the investigative reporting aspects, there are some interesting factors still meriting attention. The various parties to any investigation routinely engage in conjectural conversations. In this instance, it was universally agreed the engine failures were related to fuel problems. We also considered neither pilot possessed the requisite knowledge of the airplane's fuel system, otherwise the fuel starvations would not have occurred. Realistically, the placards required by the mandatory airworthiness directive that had not been complied with were considered to be a crutch for pilots with inadequate knowledge of the fuel system. Normally, the FAA cannot be induced to make such mandatory requirements that basically compensate for inadequately trained pilots. However, they probably did so in this instance because of an inordinate number of problems resulting from mismanagement of the fuel system in Ted Smith model 600 Aerostar airplanes.

In an entirely different ballpark, the odds of both engines failing simultaneously while being operated on separate fuel supplies are probably too remote to compute. Normally, there would be a time differential of several seconds to several minutes. Accordingly, those involved in the investigation speculated the pilots had already lost one engine due to fuel starvation, and reported the loss of both engines when the other one failed for the same reason. Knowing what we all now know about the fuel system, it probably wouldn't have taken much thought on our part before we switched the fuel valves from the X-Feed to the ON positions to gain access to the 16 gallons of fuel remaining in the fuselage tank.

At the conclusion of my testimony, I was not subject to recall. I drove back to Lambert Field in St. Louis, the location of my first training flight as a Naval aviation cadet back in 1943. Having some time to kill, and being in a nostalgic mood, a search was made for some of our old hangouts, without success. Apparently, they had given way to this thing we call progress. My first flight from Lambert Field was made in an open cockpit, two-place, yellow biplane called an N3N. It was sometimes referred to as the "Yellow Peril," a misnomer, because in reality it was pretty docile. It was equipped with tandem seating and had no radio equipment. The instructor, who normally occupied the front seat, used a gosport, which is a one-way voice communication tube, to talk to the student. The student couldn't talk back, but the instructor could use his rear view mirror to see the student shake or nod his head, or maybe use facial expressions.

Somehow, that sounds just like what most husbands need after the bloom wears off, say a week or so after the wedding ceremony. We all may have heard about the fellow who, upon being told his wife was outspoken, asked, "By whom?" Now, if he had one of those gosport contraptions and could have held onto the talking end, he'd — but no, that would be living closer to heaven than our Creator ever intended any married man to enjoy while still earthbound.

Government credentials issued to Safety Board aircraft accident investigators authorize travel on the jump seat, an extra seat provided in the cockpit of air carrier models. Such authorization is essential to keep investigators abreast of the current state of the art in air carrier operations. After my return to Miami on a jump seat, our General Counsel called to ask how matters had turned out at the trial. In reply, I told him what the judge had said. He responded: "Well, if that's what the judge said, I don't care what anyone else may have thought," gave me a "Well Done", and invited me to stop in to see him when in Washington.

OUR MAN IN SOUTH CAROLINA

While on accident standby duty in the Miami, Florida office, we were advised a light twin-engine airplane was missing on a flight to Georgetown, South Carolina. The flight had departed from Florence, South Carolina, with the pilot and one passenger on board. Standing procedures for handling missing airplanes called for the investigator to remain in the office pending further developments, and collect the information available through use of the telephone. Such actions ensured that evidence which might later prove to be significant wouldn't be lost or destroyed. This was a sound policy. The FAA routinely issues a telegraphic notice after an airplane is reported missing. It wan't unusual to learn later the airplane was undamaged and the occupants unharmed. On some occasions, pilots just failed to close their flight plans after landing. In other instances, for some reason the flights landed at a field other than the original destination.

For the flight under discussion, the occupants initially departed from Teterboro, New Jersey, with Tallahassee, Florida the ultimate destination. The pilot did not file a flight plan and his intended route of flight was not determined. The flight landed at New Bern, North Carolina, where 72 gallons of fuel was required to service the airplane to capacity. About 7:00 p.m., while still on the ground in New Bern, the pilot telephoned his parents in Georgetown, South Carolina, and notified them he would be arriving about 8:20 p.m. The New Bern Airport was uncontrolled (no control tower), and the exact departure time was not determined.

At 9:42 p.m., the pilot called the Florence flight service station and requested a heading to the Georgetown Airport. The control specialist noted on his direction finding equipment that the airplane was on a bearing of 055 degrees from Florence. This was well right of both the direct and airways routes between New Bern and Georgetown. The FAA specialist suggested the pilot might wish to proceed via Florence. The pilot accepted the alternative routing. When over Florence, the pilot reported one fuel tank had run dry, but, after switching tanks, everything appeared normal. Thereafter, an uneventful landing was executed in Florence.

At the late hour, refueling facilities were not immediately available. After a telephone call, an employee for a fixed base operator proceeded to the airport. The total fuel capacity was 120 gallons. The airplane was equipped with two main tanks of 30 gallons capacity each, and four auxiliary tanks of 15 gallons capacity each. Only 23.6 gallons were required to top all the tanks. Nineteen gallons were used to service two of the auxiliary tanks, and the remainder was pumped into the right main tank. The serviceman made a remark to the pilot about having to drive out to refuel an airplane that was almost full of gasoline. He reported the pilot said he knew he should have had plenty of fuel, but decided to top the tanks because one of the engines had cut out over the airport.

While on the ground, the pilot visited the Florence Flight Service Station and received a weather briefing. The direct distance between the Florence and Georgetown airports is only 57 nautical miles. Surface observations showed good weather prevailing over the planned route of flight with seven miles' visibility at Florence and ten miles at Charleston, the weather reporting station nearest to Georgetown. At about 10:30 p.m., the pilot telephoned his parents from Florence to say he would be taking off. He told his mother he would fly by the house and blink the landing lights to signal her to send a taxicab to the Georgetown Airport. The flight departed from Florence at 11:10 p.m. There were no communications with the flight after the departure, and the airplane failed to arrive at the destination.

When the FAA's established procedures failed to determine the whereabouts of the airplane, search efforts were initiated. Searches for missing airplanes over land areas of the United States are conducted by the Civil Air Patrol, with assistance by the military and other organizations on some occasions. The entire effort is coordinated by the U. S. Air Force's Rescue Coordination Center at Scott AFB in Illinois. To keep us abreast of the situation, liaison was conducted with personnel at Scott who reported the search was progressing routinely.

Two days later, my supervisor called me into his office and told me to go to South Carolina. "Why?" I asked, "did they find the missing airplane." "No," he replied, "Washington just called and told me to send someone to South Carolina." When asked what I was supposed to do, he told me to do anything possible, but to keep in touch until things quieted down. We learned the pilot was en route to meet with government officials in Tallahassee, Florida. Apparently, there had been some complaints of less than whole-hearted search efforts because of the pilot's minority status.

Obviously, I was dispatched to South Carolina to appease the complainers. Safety Board spokespersons responding to inquiries could then say: "Although the airplane hadn't been found, the search is on-going, and the Safety Board had an investigator on the scene." After arrival in Columbia, South Carolina, I visited the FAA's Flight Standards District Office whose supervisor was a close friend of mine. I then drove to Florence and reinterviewed the flight service station personnel, and the serviceman employed by the fixed base operator who refueled the airplane. Nothing substantive was learned during a visit to the Georgetown airport. My office chief was briefed daily as to my activities, while I monitored the ongoing search efforts through liaison with the Civil Air

Patrol and personnel at Scott AFB. However, in view of the information obtained while still in our Miami, Florida office, there just wasn't much to be done that would materially contribute to the investigation until the whereabouts of the airplane and fate of the occupants were determined.

A visit with my brother and his family who resided in the state was enjoyable. We had a pleasant round of golf in Myrtle Beach and visited historic Charleston. Thereafter, I stopped by the Marine Corps Air Station in Beaufort to visit some Marine Corps buddies. Several days after arriving in South Carolina, things apparently quieted down. At any rate, my supervisor terminated my vacation routines and had me return to the office.

Two weeks after the airplane failed to arrive at the destination, the search mission was terminated with negative results. Some idea of the magnitude of such efforts is evident from the following summary: A 5,535 square mile area was searched and researched by aircraft from the South Carolina Civil Air Patrol, U. S. Marine Corps, U. S. Air Force, and the South Carolina Law Enforcement Department. Additionally, a 196-man ground search party from the South Carolina National Guard participated. During the mission, 601 personnel, 103 aircraft, 85 vehicles, and 87 mobile and fixed radios were utilized, with the aircraft flying 229 sorties for a total of 354 flight hours.

Members of the pilot's family were appalled when they learned derogatory statements had been made concerning the conduct of the search efforts. They expressed their sincere gratitude to all the searchers, while abhorring the erroneous statements made by unauthorized personnel. The family stated they wished emphatically to disassociate themselves from those statements that were made without their knowledge or consent.

About five weeks after the search was suspended, the wreckage was found by a landowner checking timber on his property northwest of Georgetown. The airplane dove at a steep angle into the soft, swampy terrain. The extreme impact forces resulted in complete disintegration of the aircraft structure. Sighting the wreckage from the air through the dense foliage was next to impossible because the path through the trees and undergrowth was hardly noticeable. The engines were found buried five feet beneath the water-filled impact crater. Local residents in the sparsely populated area around the crash site reported they may have heard the impact around midnight, or shortly thereafter, on the night of the flight.

This was another of those occurrences where it was impossible to pinpoint the precise cause. There was no reason to suspect weather involvement. The circumstances of the accident made meaningful post-mortem studies impossible. The degree of wreckage disintegration precluded a determination as to whether a failure or malfunction of the aircraft structure, flight control system, or powerplants was involved. It should be understood that even a double engine failure would not cause the airplane to dive into the swamp at such a steep angle and high velocity. Even without power, the pilot could have traded altitude for airspeed while executing a controlled descent at a much slower speed.

There was some evidence that the accident may have occurred 50 to 55 minutes after the takeoff. This did not correlate with the normal cruise speed of the airplane in conjunction with the 57-mile distance between Florence and Georgetown. On the flight out of New Bern, the non-instrument-rated pilot exhibited poor navigational practices. Additionally, he apparently mismanaged the fuel controls in a manner that caused one of the engines to fail over Florence. Such matters do not reveal a high degree of familiarity with the airplane, or the required degree of navigational proficiency by the pilot. On the dark night, the pilot may have become disorientated while flying over an area of sparse lighting on the ground. Another possibility is that the pilot became confused, and ultimately lost control of the airplane, while attempting to cope with another pilot-induced engine malfunction.

TOOLS OF THE TRADE

The discussions below relate to the basic flight controls installed on aircraft.

ELEVATORS - The elevators are the movable surfaces on the rear of the horizontal tail. They do the same thing the elevator in a building does, take you up or down. This is accomplished by controlling the position of the nose in relation to the horizon. A pilot pulls back on the control yoke or stick to raise the nose, and pushes forward to lower the nose. The throttles are used in conjunction with the elevators to maintain an appropriate airspeed and the desired rate of climb or descent.

Most of us have seen films showing bombers returning to England during WW II with many components shot off. However, elevator control is essential to controlled flight, and few, if any, of those bombers returned when both elevators, one on each side of the horizontal tail, were missing.

Examples where pilots or aircrews have coped, to any degree of success, with a complete loss of elevator control are rare. However, such wonderous happenings do occur. For instance, the flight controls of the McDonnell-Douglas DC-10, and some other jet models, are connected to hydraulic actuators that actually move the control surfaces. In such installations, a complete loss of hydraulic pressure results in inoperable flight controls. This is what happened when the McDonnell-Douglas DC-10 crashed in Sioux City, Iowa. The pilots lost all elevator control when disintegration of the tail engine's rotor fan disk destroyed some hydraulic components of the flight control system. It was truly a remarkable accomplishment when the flightcrew managed to cope with the consequences of the complete loss of elevator control to a degree that permitted 184 of the occupants to survive the accident.

RUDDER - The rudder is the movable surface attached to the rear of the vertical fin on the tail of the airplane. It provides directional control which can best be understood by relating it to the steering wheel in your car or the rudder on a boat. The pilot uses right rudder to turn right, etc. When turning to a new heading, coordinated flight is maintained through simultaneous use of the rudder and ailerons.

AILERONS - The ailerons are the movable surfaces on the outboard trailing edges of the wings. Basically, they control the bank angle of the wings. However, it is possible to maintain some control over the wing bank angle through use of the rudder, and to maintain some degree of directional control through use of the ailerons. For this reason, some of those WW II bombers were able to return to England with nearly the entire vertical tail, or very little of the aileron surfaces, remaining.

TRIM TABS - The elevator, rudder, and aileron control surfaces on many airplanes have trim tabs the pilot may utilize to permit near hands-off flight. One of my investigations involved a pilot who managed a safe landing after losing elevator control. The airplane had recently been returned to service after receiving a new coat of paint. The elevators were removed and reinstalled during the paint operation. After taking off, the pilot discovered he had no elevator control. However, the elevator trim tab was still operational. He remained airborne until gaining proficiency at maintaining control through use of the trim tab. Thereafter the pilot executed a safe landing. Investigation showed the elevator cable had come loose in the tail and the attaching hardware was found in the aft fuselage belly. Although the mechanic denied any involvement, it was apparent he had failed to properly reassemble the elevator components.

WING FLAPS - Wing flaps are attached to the inboard trailing edges of the wings and sometimes extend beneath the fuselage. Basically, they deploy downward, to a degree selected by the pilot, to alter the shape of the wing so it will generate greater lift at the slower airspeeds during takeoffs and landings. The benefits derived through use of the wing flaps, and the disastrous consequences that can occur when their use is neglected, can be gleaned from the accident at Detroit, Michigan on August 6, 1987. For takeoff on a flight to Phoenix, Arizona, the Northwest Airlines flightcrew failed to extend the wing flaps on the McDonnell-Douglas MD-80 jetliner. The stall warning activated shortly after liftoff and the airplane crashed, killing 157 passengers and crewmembers. Miraculously, a four-year-old girl passenger survived the tragedy.

Although not a recommended procedure, the airplane would have been capable of taking off with the wing flaps retracted. On such a takeoff, the airplane would have had to be accelerated to a much higher airspeed before being rotated for liftoff. In the accident, the flightcrew, assuming the wing flaps were extended, lifted off at the predetermined rotation speed. The stall warning activated immediately because the airplane had not attained an appropriate airspeed for the takeoff configuration.

WING SLATS - Wing slats are movable auxiliary airfoils running along the leading edge of the wings of most commercial jet models. When deployed, they move out and down, usually in conjunction with deployment of the wing flaps. They also reshape the wing so it will generate greater lift at the slower airspeeds prevalent during takeoffs and landings.

Uncommanded retraction of the leading edge slats on the outboard portion of the left wing was a primary factor in the American Airlines, McDonnell-Douglas DC-10 accident at Chicago, Illinois in 1979 that killed 273 people. The crash occurred when the engine installed on the left wing fell off the airplane shortly after the takeoff.

Other damage to the left wing associated with separation of the engine resulted in retraction of the outboard wing slats. Retraction of the wing slats at the relatively low takeoff speed caused the left wing to stall. Thereafter, the airplane rolled over to the left before impacting the ground about a mile beyond the departure end of the runway.

SPOILERS - Spoilers are tabs, usually smaller than the wing flaps or slats, that move up on the wing to deflect the airstream away from the upper wing surface. On some aircraft, they are authorized for use in flight to slow the airspeed rapidly, or greatly increase the rate of descent. In reality, they create conditions somewhat similar to those occurring when an airplane stalls.

The angle of attack is the relationship of the wing to the relative wind. Stalls occur when the angle of attack gets too steep. In lay terminology, an airplane stalls when in too much of a nose-high attitude. Under such conditions, the airstream is unable to turn the somewhat sharp corner over the wing leading edge and still adhere to the upper surface of the wing coming back down. Obviously, when in a vertical, high-speed climb, the wing is streamlined into the "relative" wind until the speed diminishes. Thereafter, the airplane stalls. During airshows, stunt pilots demonstrate the condition by executing a "hammerhead" stall at the top. That's where the airplane abruptly transitions from a vertical climb, to a vertical dive, before the pilot recovers.

Ground spoilers, that may or may not be the same as those utilized in flight, also extend to deflect the airstream away from the upper surface of the wing. Their use shortly after touchdown effectively kills the lifting force generated by the wings. This causes the weight of the airplane to settle rapidly onto the wheels, which greatly increases braking action on the landing rollout.

Next time you are on a jet airplane, keep your eye on the upper surface of the wing and you will see the spoilers extend upward immediately, or shortly after, touchdown.

Three crewmembers died when a Grumman G-1159 Corporate jet, owned and operated by International Business Machines, Inc., crashed on a training flight. The Safety Board's probable cause determination showed unwanted extension of the ground spoilers to be the predominating factor in the accident.

COCKPIT CONTROLS - Insofar as the pilot is concerned, the cockpit controls are basically the same regardless of the methods used to transmit pilot inputs to the flight control surfaces. For many years, airplane flight control systems used flexible, stranded wire cables, or rigid, push-pull rods, with direct connections from the cockpit controls to the movable control surfaces on the wings and tail. However, this is not true for some of our newer military jet fighters or the latest Airbus Industrie A-300 series, twin-engine, fan-jet, transport airplanes being manufactured in France. For this discussion, it is sufficient to say that those airplanes employ a sophisticated fly-by-wire concept using electrical signal inputs from side-arm controllers in the cockpit.

HELICOPTERS - The main rotor blades on a helicopter are simply rotating airfoils that allow it to lift off vertically at zero airspeed. The centrifugal motion of the rotor blades through the air produces the lifting force. Remember, it was Sir Isaac Newton, the English mathematician and philosopher, who figured out the laws of motion. Basically, he found that for every action, there is an equal and opposite reaction. So, in accordance with Sir Isaac's laws, as the main rotor turns, the fuselage wants to rotate in the opposite direction. This is called torque. The vertical propeller on the tail is an antitorque rotor that also provides directional control. Changes in engine power result in a corresponding change in the torque effect on the fuselage. The foot pedals in the cockpit are linked to the pitch change mechanism in the tail rotor gear box. They are used by the pilot to increase or decrease tail rotor thrust, as necessary, to maintain directional control and neutralize the torque effect.

Loss of tail rotor effectiveness does not necessarily result in an accident. On many occasions, pilots have landed safely under such conditions. In one instance a helicopter pilot on a fish-spotting mission lost all tail-rotor control. Although it required several attempts, he managed a safe landing on the rather restricted confines of the landing platform on the ship. However, if a helicopter is not landed safely after a loss of tail rotor control, the Safety Board does not normally show pilot factors in its probable cause determination.

A false balance is abomination to the Lord: but a just weight is his delight.

Proverbs 11, Verse 1.

WEIGHT WATCHING

King Solomon's saying above probably had reference to the honesty of merchants, vendors, clerks, etc. However, in modern times, his words constitute sage advice for pilots when it comes to the matter of properly loading their airplanes. It is an aspect of flying to which pilots sometimes give only casual consideration. In other instances, there are pilots who exhibit almost total indifference towards their responsibility to ensure that their airplanes are properly loaded so as to comply with stipulated weight and balance limitations. And we are not referencing only inexperienced pilots. Those with a significant number of flight hours are sometimes guilty of violating the weight and/or the balance limitations for their airplanes. In many instances, an accident was inevitable once the pilot attempted to take off with the weight, or the balance, outside the prescribed limits.

Readers are encouraged to acquire an understanding of the following definitions before continuing. A basic understanding of these terms should ensure that everyone will be capable of comprehending the factors involved in the accidents to be reviewed.

GROSS WEIGHT - The gross weight includes the airplane, fuel, oil, cargo, pilot, passengers, etc. In other words, it includes everything that weighs anything. The FAA requires that manufacturers stipulate the maximum authorized gross weight for each of their airplane models.

BALANCE - Balance relates to an airplane's center of gravity. The center of gravity is the point about which an aircraft would balance (hang in a level attitude) if it were possible to suspend it at that point.

REFERENCE DATUM - Simply stated, a fixed point to measure from. Manufacturers must establish a reference datum for each of their airplane models. It makes little difference where the reference datum is located; however, once established, all arms, and the locations of the center of gravity and the permissible center of gravity range, must be measured from that point.

ARM - Arm is the distance in inches from the reference datum to the location of an item in the fuselage. (Pilot, passenger, fuel, cargo, etc.)

MOMENT - Moment is the product of the weight of an item in pounds, multiplied by its arm in inches.

CENTER OF GRAVITY - Once the total weights and moments for the airplane and everything on it are calculated, the location of the center of gravity is determined by dividing the total moment by the total weight. It is expressed in inches aft of the reference datum.

CENTER OF GRAVITY RANGE - The permissible center of gravity range is the distance between the authorized forward center of gravity location (expressed in inches aft of the reference datum), and the authorized aft center of gravity location. The FAA requires that manufacturers provide the center of gravity range for each of their airplane models. Strict compliance with the center of gravity limitations is essential to assure that the airplane will be controllable during all authorized phases of flight.

A simple balance problem is a 200-pound grandfather seesawing with his 50-pound grandson. If the grandson is seated 120 inches from the center support, his moment is 6,000 inch-pounds (50 X 120). If the grandfather wishes to provide perfect balance at the crossbar, he would have to sit 30 inches from the center support to provide the required equal moment of 6,000 inch-pounds (200 X 30).

It is possible to overload an airplane with too much weight and still have it within the authorized balance range. Conversely, it is possible to load it within the weight limitations, but have the weight distributed in a manner that results in a center of gravity that is outside the authorized range. Additionally, the pilot must understand that in many aircraft it is not possible to fill all seats, load the maximum baggage, top off the fuel tanks, and still remain within the stipulated weight limitation. Often, the pilot must reduce the baggage weight, or the fuel load, or both, in order to carry a passenger in every seat.

Excessive weight produces deficiencies in almost every aspect of an aircraft's performance. Some of the more important deficiencies resulting from overweight conditions are: higher takeoff speed; longer takeoff run; reduced cruising speed; higher stalling speed; higher landing speed; longer landing roll; etc. Although excessive weight is never to be condoned, a slight overweight condition is probably preferable to a center of gravity location that is outside the stipulated range to any degree whatsoever.

It is possible to have uneven balance conditions in the lateral axis, meaning from wingtip to wingtip. However, the design of the aircraft usually ensures lateral balance unless the pilot has unequal fuel quantities in the wing tanks. This normally results only in wing heaviness on one side or the other, and such a condition is rarely uncontrollable. Accordingly, the following discussions relate to balance conditions on the longitudinal axis, or from the nose to the tail.

In most accidents caused by improper balance conditions, the center of gravity exceeds the aft limit, producing a tail-heavy condition. However, on occasion, an accident occurs because the center of gravity exceeds the forward limit, which produces a nose-heavy condition.

Remember, in our discussion of flight control systems, we learned that the elevators, which are the movable surfaces on the horizontal tail, provide longitudinal, or fore and aft control. Pilots use the elevator flight control to raise or to lower the nose. In cases where the airplane is loaded so as to provide a center of gravity forward of the forward limit, he may have insufficient elevator control to raise the nose. Conversely, if the center of gravity is located aft of the rear limit, the pilot may have insufficient elevator control to lower the nose. And as we shall learn, either event can be a recipe for disaster.

Pilots are provided with the weight, arm, and the center of gravity for the empty airplane. Thereafter, the center of gravity will vary, depending upon the precise manner in which the pilot distributes the load. This might require that the pilot have passengers occupy specific seats, while possibly limiting the fuel quantity, and otherwise distributing the baggage or cargo in a manner that provides a gross weight and center of gravity that fall within the prescribed limits. Once pilots decide how their airplanes are to be loaded, they are able to compute the total weight and the total moment. Then, to determine the center of gravity, they divide the total moment by the total weight.

Pilots are not left with their individual mathematical skills when it comes to determining the gross weight and center of gravity for their airplanes. The aircraft manufacturer and the FAA have major roles in designing and certificating aircraft with safe and workable methods of controlling the weight and balance. Simple and orderly procedures based on sound principles have been devised. These include charts, graphs, and sample problems with detailed explanations concerning their use. Pilots who have studied the weight and balance instructions for their airplanes, and who comply with the stipulated limitations, should encounter no weight and balance problems.

- -

WAY OVER AND WAY OUT

A single-engine, air-taxi, cargo flight crashed in a residential area shortly after an early morning takeoff. An entire 1,944-pound cargo shipment had been loaded aboard the airplane, producing a gross weight at takeoff that was 790 pounds over the authorized maximum weight of 3,800 pounds. The stipulated center of gravity range for the airplane at the maximum authorized weight was only 7.5 inches. It is not unusual for small, single-engine airplanes to have such a narrow center of gravity range at the maximum gross weights. The pilot was killed, and the wreckage was extensively burned in the postcrash fire. Tie-down straps had not been used to secure the cargo and some was scattered outside the cabin. This made it impossible to determine where it had been located in the cabin. Balance computations based upon the assumption that the cargo had been evenly distributed in the cabin showed the center of gravity would have been 7.5 inches aft of most rearward limit at the time of the accident.

The examination of the wreckage produced no evidence to show that a failure or malfunction of the aircraft structure, flight control system, or powerplant was involved in the accident. The engine firewall prevented major fire damage to the powerplant. Examination of the engine showed no evidence that a loss of power had occurred. This correlated with the statements of most of the witnesses who noticed nothing unusual in the operation of the engine before the crash.

The circumstances of the accident, showing almost total disregard for the stipulated weight and balance limitations, were appalling to both myself and the two FAA inspectors participating in the investigation. There was little doubt in our minds that weight and balance factors were involved in the cause of the accident. Accordingly, our initial action after completing the on-scene investigation and making arrangements to have the wreckage moved so the neighborhood could return to some degree of normality, was to visit the operator's headquarters. In view of the evidence, we very much wanted to determine the details associated with such flagrant violations of the weight and balance limitations.

We met with the Air-Taxi Operator's President who described his arrangements for the flight while providing us with a written statement of the circumstances. The substance of his statement was as follows: "We received a shipment of freight totaling 1,941 pounds. We decided the best way to ship that much freight would be to cut an extra section and send half the freight that night. The remainder would be shipped aboard our regularly scheduled flight at 5:30 a.m. in the morning. I loaded the freight myself with help from one of my employees. I left 700/800 pounds in the truck with an airbill to be delivered at 5:30 a.m. the next day. I called the pilot and explained how this was to be handled. He was to use the airplane for two trips, and since we only had 41 pounds of freight for the regular 5:30 a.m. flight, this should have worked out well."

Investigation showed the truck with the cargo to be delivered on the 5:30 a.m. flight was left parked in front of the operator's office. Another company pilot, making plans for a flight to a different destination, said he noticed the truck was brought around from the front of the office and backed up to the side of the airplane involved in the accident. He was unable to provide additional information, stating he was busy with his flight preparations and not particularly observant of the actions of the other pilot. All this was happening at about 1:30 a.m., and we were unable to locate witnesses who could provide additional information.

Company records showed the well qualified, 23-year-old pilot had 2,750 flight hours, including 1,600 hours in single-engine models. It was considered we had a pretty firm handle on the cause of the accident and I was busy readying my investigative report. Before mailing it to Washington, I received three letters from the pilot's father, and another from a friend, who had been the pilot's flight instructor, imploring that we dig deeper into the circumstances connected with the loading of the entire cargo shipment on the flight. They claimed only 16 minutes elapsed between the time the pilot arrived at the airport and his takeoff. Basically, they contended that provided insufficient time for the pilot to file his round-robin flight plan, move the truck from the front of the office to the rear, and load the remainder of the cargo on the flight. They forwarded the results of time-motion studies they had conducted to back up their contentions.

Their requests presented us with somewhat of a paradoxical dilemma. The employee who had assisted the company's chief executive officer with the cargo loading said they only put part of the cargo on the airplane. Whenever possible, attempts were made to acquiesce to the wishes and desires of the next-of-kin of those killed in the accidents I investigated. Oftentimes, they had little or no knowledge of aviation, and I figured it was my duty to enlighten them whenever possible. Accordingly, I was prone to provide more information than called for in the Safety Board's guidelines, while also attempting to explain enough about the particular aspects of aviation involved so they would understand what we were talking about. However, in such instances, it is sometimes difficult to make people understand that the ultimate authority, and the individual burdened with the sole responsibility for the safe conduct of any flight, rests with the pilot-in-command.

We considered a satisfactory inquiry had been conducted for the Safety Board to fulfill its congressional mandate to thoroughly investigate the accident, determine the factors involved, publish the pertinent evidence, and make its probable cause determinations. We had also obtained some information relating to the matters discussed by the petitioners. It was considered efforts to delve more deeply into these matters would exceed the Safety Board's area of responsibility. In hopes of finding a resolution to the question, the entire file was taken into my supervisor's office. After studying the pertinent data and discussing the evidence obtained during the investigation, he said: "Tom, what they're asking you to do is more in the civil arena involving culpability. I suggest you allude to their requests in the narrative section of your report, and then attach their letters to the accident file. Such actions will preserve everything, and show complete impartiality, while otherwise fulfilling our obligations."

I reasoned that was why we have supervisors, and his suggestions were carried out. However, I am aware such actions fell far short of satisfying the pilot's father, and to this day, that is still bothersome.

TOO NOSE HEAVY

"This is a very interesting accident insofar as the weight and balance aspects are concerned. It is an excellent example of the manner in which weight and balance factors can become

involved regardless of the experience level of the pilot.

The above quotation is from a memorandum attached to the report of an accident involving two fatalities we were forwarding to Washington, D.C. It involved a single-engine, tailwheel-equipped airplane flown by a 55-year-old pilot who had about 15,000 flight hours, including more than 1,000 hours in light airplanes. He possessed ratings in large, turbojet-powered transport models, and was employed as an airline captain by one of our major air carriers. His 22-year-old son, a State University student and certificated private pilot with 258 flight hours, occupied the right front seat.

The accident occurred at the pilot's private airstrip adjacent to his residence. Investigation showed the airplane collided with trees about a mile north of the airstrip while the pilots were attempting a go-around from a landing approach. The 2,600-foot grass strip was aligned north and south. There were no obstructions off the southern end of the strip. However, trees approximately 100 feet tall were just beyond the northern end. The pilot's spouse said her husband and son always took off to the south and landed to the north.

The examination of the wreckage showed no evidence of preimpact failure or malfunction of the aircraft structure, flight control system, or powerplant. There were no eyewitnesses to the accident, but statements were received from witnesses who had seen the airplane flying in the area and heard the engine rev up to a high power setting just before the crash. Propeller slash marks on trees at the accident site showed the engine was developing significant power at the time of the crash, and the throttle was found in the wide open position in the wreckage.

Investigation showed the airplane was 260 pounds under the maximum authorized gross weight of 3,350 pounds at the time of the accident. However, balance computations showed the center of gravity was 1.4 inches forward of the authorized limit. In this instance, there was no baggage or cargo on the airplane, just a full fuel load, the 190-pound pilot and his 343-pound son. It was determined the pilot and his son had flown the airplane under similar conditions many times in the past. Accordingly, it might be concluded that the center of gravity location, although exceeding the forward limit, was not involved in the accident.

However, that would be an erroneous conclusion. The accident was related to the airspeed maintained by the pilots in conjunction with the balance condition that was outside the authorized range. In any airplane, a larger deflection of the elevator is required to raise the nose as the airspeed decreases. Readers should understand this is something pilots who operate their airplanes within the authorized weight and balance limitations never have to worry about. That is because the FAA's airworthiness standards require that airplanes be controllable throughout the authorized flight envelope before they can be certificated.

The circumstances of the accident, in conjunction with the observations of the witnesses, showed the pilots were executing a landing approach to the north. For undetermined reasons, they abandoned the landing approach and commenced a go-around. Remember, the pilot's spouse said they always took off to the south in order to avoid having to climb over the trees at the northern end of the strip. Although the witnesses heard the engine rev up to a high power setting, and the throttle was found in the wide open position, it was reasonable to assume the pilots were forced to operate at a very slow airspeed in their attempt to clear the trees. Thereafter, the slow airspeed, in conjunction with the center of gravity location forward of the authorized limit, resulted in there being insufficient elevator control to prevent the airplane from crashing into the treetops.

Investigators exhibit a keen interest while collecting the relevant facts, circumstances, and conditions of such perplexing occurrences. We put forth conscientious efforts in our endeavors to determine the manner by which such qualified pilots could be caught off-guard. Casual onlookers or listeners might conclude we were without an appropriate degree of compassion for those who perished in the accident or for their next-of-kin. However, that is not the case at all. We merely become so completely engrossed when analyzing the evidence and conjuring up all manner of theoretical probabilities that we sometimes sound insensitive.

In this instance, we considered we were on fairly firm ground when we concluded the occupants had never before attempted flight at such a slow airspeed when so close to the ground. Additionally, it was reasonable to assume they had previously operated near the stall speed while flying at a safe altitude. However, slow flight at a safe altitude does not necessarily demonstrate the dire consequences of maneuvering at such slow speeds when flying close to the ground. If the nose fell through because of insufficient elevator control at a safe altitude, the pilots would only be required to continue the descent until accelerating to an airspeed high enough to regain sufficient elevator effectiveness to raise the nose. Don't get lost in this arena. Remember, we said the slow airspeed attempted by the pilots in conjunction with the fact that they had no excess altitude to trade for a higher airspeed caused the accident.

We've all heard the sayings: "For want of a nail, the shoe was lost, for want of a shoe, the horse was lost, etc." Well, concerning this mishap, we might borrow the theme and say: "For want of 50 pounds of dead-weight in the baggage compartment, the plane was lost." This is true because that much weight in the baggage compartment behind the passenger seats would have moved the center of gravity two inches further aft. That would have put it in the permissible range where the FAA requires sufficient elevator effectiveness to ensure longitudinal, or fore and aft control, from the stall speed, to the maximum authorized speed for the airplane.

- -

CORNER CUTTING

In many fields of endeavor, safety can be jeopardized by noncompliance with regulations, operation of defective equipment, utilization of unqualified personnel, or other actions that might result in a tainted financial return. We have to understand, aviation has its unscrupulous operators. On some occasions, even reputable operators have been caught cutting corners. When we add all those who make it a practice to operate on the fringes of the regulations, plus those who habitually disregard the regulations whenever possible, it becomes apparent it would take a whole army of FAA inspectors to uncover every shady practice or other unsafe circumstance brought about by operators who fail to comply with the regulations or other established procedures. The entire system is supposed to operate on the premise that operators are going to strive to comply with the regulations while employing sound operating and maintenance procedures.

On many occasions, it was my opinion the Safety Board unjustly cited the FAA for failing to provide adequate surveillance of operators in the finding and conclusion portions of its accident reports. Many such findings were thought to fall into the "Monday morning quarterbacking arena." The FAA will never have enough inspectors to provide the degree of surveillance necessary to uncover all infractions. The FAA regulations set forth the minimum standards. Thereafter, the aircraft, engine, and component manufacturers, working in close harmony with the various operators, are expected to utilize the FAA standards as a basic guideline when establishing sound operating procedures, adequate inspection programs, and proper maintenance criteria.

Although providing little solace to accident victims, there is one redeeming feature when airplanes belonging to shady operators are involved in an accident. It is derived from the fact that Congress has given the Safety Board the authority it needs to properly carry out its investigations. Some operators manage to skirt the regulations or otherwise cut corners until their unsafe practices result in an accident. Thereafter, their unscrupulous activity is exposed. It is not unusual for them to belittle the investigative efforts, or try to point out someone or something else, as they attempt to divert the thrust of the investigation away from the actual factors involved. However, they rarely succeed and their shady practices are invariably brought to light. The occurrence discussed below might fall into that category.

- -

UNCOOPERATIVE

The trusty old twin-engine, Douglas DC-3, tailwheel-equipped airplane crashed while taking off on a flight from Miami, Florida to Havana, Cuba. It happened after Castro, who probably figured out a way to get his hands on some hard currency out of the arrangement, allowed visits by some Cuban Americans with their relatives. There were no serious injuries to the 24 passengers or three crewmembers. However, the airplane sustained substantial damage when it crashed in a grass area off the right side of the runway.

Our office was nearby, and upon my arrival at the scene, a pickup truck loaded with baggage was about to pull away from the aft baggage compartment. When asked where the baggage came from, the driver said he had removed it from the aft baggage compartment. I asked, "All of it?" And he replied, "Yes." The driver was anxious to leave, saying the passengers were waiting at the terminal to collect their bags. After showing him my credentials, he was instructed to delay until we could arrange to have the baggage weighed.

The takeoff was made behind a landing Boeing 737, twin-engine commercial jet. The tower operator cleared the flight into position with instructions to hold until the landing traffic had cleared the runway. The runway was on the south side of the airport, several hundred feet inboard of the perimeter fence. A 50-year-old officer in the Civil Air Patrol with about 150 hours flight experience observed the takeoff from a position just outside the fence. He was located adjacent to the point where the wreckage came to rest. He said the airplane lifted off in a three-point attitude after a short takeoff roll and rose almost vertically to a height of about 50 feet before it began descending. At the end of the statement regarding his observations, he provided the following analysis:

> **"The one very distinct fact that struck me as I observed the aircraft rotate was that, in my opinion, it did not achieve anything near to the ground speed it should have. The tail gear did not elevate at anytime during the whole episode until the aircraft nosed over. It appeared to be a departure stall, caused by too short a takeoff roll, too little ground speed, and perhaps, an aft center of gravity.**

It was my policy to be an attentive listener when witnesses wanted to explain the cause of an accident. Some wild-haired guesses were heard, but I also listened to some pretty close calls. Upon returning to the office, I remarked to my supervisor we may have found a likely candidate for our vacant investigator slot. I told him the candidate did not possess extensive flight experience; however, he had come up with some rather astute observations concerning the factors involved in the crash I was investigating.

The airplane impacted the ground tailwheel first, just off the right side, and about 3,000 feet down the runway. It dragged the tailwheel through the terrain a distance of 60 feet before the main landing gear wheels impacted the ground. Thereafter, the landing gears collapsed into the wheelwells, and the landing gear doors, the oil coolers that were attached to the bottom of the engines, and the right propeller separated before the wreckage slid to a stop in an upright attitude.

The passengers deplaned through the aft cabin door. A bus was dispatched and they were returned to the terminal with their carry-on baggage before my arrival. After making a cursory examination of the crash site, talking to the witnesses, and being apprised of the circumstances of the accident, we very much wanted to know the weight of the passengers' carry-on baggage. We were suspicious, because passengers travelling to locations where so many of our daily necessities are not available oftentimes carry some fairly weighty items. However, we learned most of the passengers had already departed the terminal and it would have been next to impossible to reassemble them for a weigh-in.

The baggage the passengers checked through to the destination was weighed when they initially arrived at the airline's passenger service counter in the terminal. As required for international flights, the flightcrew included a copy of the passenger manifest in the general declaration documents they filed with U. S. Customs. A copy of the passenger manifest obtained from the customs office showed the checked baggage weighed 925 pounds. That was accurate, because we came up with a weight of 924 pounds for the baggage removed from the aft baggage compartment. However, instead of the actual 925 pounds, the flightcrew used a total baggage weight of only 500 pounds in their weight and balance computations.

The maximum authorized gross weight for the airplane was 26,200 pounds. The aft baggage compartment had a warning placard showing a maximum authorized weight of 600 pounds. Recomputation of the weight to allow for the additional 424 pounds of baggage placed in the aft baggage compartment produced a takeoff gross weight of 26,098 pounds. Recomputation of the center of gravity showed it was slightly less than an inch forward of the aft limit. A review of records for a similar flight to Cuba in the airplane three days earlier showed that flightcrew also used 500 pounds for the total baggage weight. A copy of the passenger manifest that was obtained from the customs office showed the baggage actually weighed 911 pounds.

It was obvious the operator was giving only lip service to weight and balance matters. The witness observations, in conjunction with the evidence we had obtained in our investigation, showed the accident occurred because of a premature liftoff at a relatively slow airspeed. The aft center of gravity condition, coupled with delayed reactions by the flightcrew, were considered to be the primary factors associated with the abrupt climb after the liftoff. Although very damaging to the flightcrew, the fact that the tailwheel dug through the terrain 60 feet before the main landing gear wheels ever impacted the ground, supported our contentions. In other words, the evidence accurately meshed with the circumstances. We figured the airplane was actually a little overweight, and the center of gravity was probably located slightly aft of the rear limit because of the suspected weight of the carry-on baggage.

The crux of the matter came down to the takeoff technique used by the flightcrew. During a normal takeoff in the Douglas DC-3 airplane, it is allowed to accelerate down the runway in a three-point attitude until reaching a speed sufficient to raise the tail. After the pilot lifts the tail through use of the elevator control, the airplane continues to accelerate until attaining the desired liftoff speed. Although not considered to have been involved in the accident, it was apparent the pilot failed to consider the hazardous conditions that could have been encountered because of the wake turbulence created by the landing Boeing 737 airplane. Sound operating procedures required that the pilot delay the liftoff until past the point on the runway where the jet had landed. The runway length was 9,350 feet, far more than would ever be required to accelerate the airplane to an appropriate takeoff speed.

The principal officer for the airline company was employed as a captain on commercial jets by one of our major airlines. He steadfastly refused to give any significance whatsoever to our findings concerning the weight and balance. He discarded the evidence with the comment: "Everyone knows the DC-3 airplane is controllable when slightly out of the weight and balance limitations." In an obvious attempt to point his finger at somebody else, he insisted the thrust of the investigation should be directed towards a study of the wake turbulence effects from the Boeing 737 that landed and taxied off the runway before their flight took off. However, that was irrelevant. It is the pilot's responsibility to alter the takeoff or landing point, or the flight path, so as to avoid the wake turbulence produced by jet transports.

The pilots were even less cooperative, and would not let us interview them. Instead of assisting with the investigation, company personnel turned the matter over to their attorney. That presented no problems; we realized the flightcrew had the right to be represented by counsel and we routinely conducted many interviews under such circumstances. However, in this instance, the attorney refused to let us interview the flightcrew members. He stated during a telephone conversation that, after preliminary discussions with the flightcrew, he "didn't feel comfortable," leaving me to determine what that meant.

He next attempted to have me provide him with a copy of my entire investigative file. With the complete backing of my supervisor, we steadfastly refused, although he was provided some data. He persisted in his attempt to gain access to my file by writing letters to our Washington staff; however, I was never directed to give him everything in my file. In a letter dated two days after the accident, the attorney said it was his intention that the crew cooperate with the Safety Board in its investigation. Ten days after the accident, we received the operator's accident report that had statements by the flight crewmembers attached. After reviewing their statements, we determined a further interview would not be required.

The crewmember statements provided the details of the occurrence but contained nothing to discredit the evidence we had obtained. The operator's uncooperative attitude, in conjunction with the refusal of the principal officer of the airline to attach any significance whatsoever to the shoddy operational practices we had uncovered, prompted me to make the following recommendation proposal:

The operator's procedures to ensure proper weight and balance control are

inadequate, and place the safety of the public in jeopardy. The attitude of the company's chief executive officer, and his reluctance to attach any significance whatsoever to the investigative findings, shows little or no corrective action will be taken voluntarily. Accordingly, emergency revocation of the operator's air carrier certificate is recommended

The recommendation proposal was not processed by our Washington staff into a formal recommendation to the FAA. However, under similar circumstances today, I would make the same proposal.

- -

A WELL PLANNED MISHAP

A temporary auxiliary fuel system was installed in the aerial application (crop dusting) airplane using the 278 gallon hopper for fuel storage. Such modifications constituted a major aspect of the operator's business. The modifications were made in preparation for a ferry flight to Venezuela.

The pilot took on a full load of fuel before departing from the operator's base in central Florida. That made the airplane ten percent over the normal gross weight. The excess weight was authorized by the FAA, but only for the ferry flight.

After taking off, the pilot commenced making practice landings. Several aviation-oriented witnesses became quite concerned for the safety of the flight as they noted the high power settings required in conjunction with the nose-high attitude and slow airspeed as the airplane struggled around the airfield. One of the witnesses was the Director of Advanced Engineering for an aircraft manufacturer with over 30 years' experience in all aspects of light airplane design. While having lunch in the airport restaurant, he observed the airplane making the circuits around the field. His statement included the following:

> **On the final takeoff, the flight was airborne just before the midfield intersection with slow airspeed, a nose high attitude, and a low climb rate. I was sufficiently concerned at the marginal appearance of the preceding climbouts that I watched this one more closely. The airplane continued its slow climb, then entered a gentle left turn, still nose high and still climbing, although "mushing" slightly early in the turn. The slow climbing turn increased my concern to the extent that I literally thought: "How much of this CAN you expect to get away with?"**

The witness did not have a long wait before getting an answer to his thoughts. A few moments later, the airplane rolled to a near vertical wing attitude, the nose dropped through, and the airplane entered a left tail spin into the ground. The pilot was killed and the airplane was destroyed.

It is believed everyone will recognize that the pilot's failure to maintain a safe flying speed caused the accident. However, the exercise of such poor judgment on his part presented the really puzzling aspect. If he intended to shoot landings, why did he take on so much fuel? Neither the operator nor any of his employees had any inkling the pilot intended to do anything other than proceed on the ferry flight. This is an example of yet another needless, tragic accident.

TROUBLE BREW RECIPE - Mix together equal parts of complacency and over confidence, whip moderately with pilot fatigue, spice generously with marginal weather, and garnish with a goodly portion of get-home-itis.

IN SEARCH OF PERFECTION

Experience is recognized as being a great teacher. However, accidents happen to some of the best qualified individuals in any field of endeavor. Pilots must guard against their lofty flight experience levels leading to two of its pitfalls, overconfidence and complacency. Or, as some might say, "become faultless to a fault." In his book "THE LEFT SEAT," author Robert J. Serling relates the following testimony by a witness during the public hearing phase of the Board's investigation of a catastrophic air carrier accident:

"I should like to point out to the Board if I may, that Captain (name omitted) had more than eighteen thousand hours in the air. As his immediate superior and supervisor, it is incomprehensible to me that a pilot with such experience would descend blindly into known mountainous terrain."

Without intending to discredit his testimony in any manner whatsoever, it may not have been wise to wager the family bankroll on his assertion. The journeyman accident investigator has learned to expect no surprises. Investigators are aware that a miscalculation, error in judgment, lapse of memory, etc., are always a possibility.

- -

IMPRESSIVE CREDENTIALS

The former U. S. Naval Aviator and Air Force General had been presented several prestigious awards for his achievements in aviation. He was one of the pioneers in the utilization of flying boats for passenger service. He also developed and used aerial navigation procedures for the polar regions before all the modern-day aids came upon the scene. He was the president of a commuter airline operating amphibious airplanes when I first met him. At that time, I was aware he had been required by the FAA to accept mandatory retirement from duties as an airline captain upon reaching age 60. Glancing at his youthful appearing, lean, tall, and handsome frame, I asked where he had found the fountain of youth. He laughed, and replied he got a little exercise and watched what he ate.

The airline's president had considerable political influence. People possessing such clout are sometimes capable of swaying governmental actions. Such was the case in this instance insofar as the surveillance activities of the responsible FAA office were concerned. This was exemplified in the Safety Board's findings for an accident on one of the airline's flights being piloted by their famous president. I was not involved in the investigation. The Safety Board assigned an Investigator-in-Charge from its Washington headquarters and used the team concept for its inquiry. The following was exerpted from the Safety Board's news release following its determination of the probable cause for the accident:

> **A commuter airline amphibian crashed off St. Thomas in the Virgin Islands, when one of the aircraft's engines failed. The crash of the twin-engine Grumman Goose operated by Antilles Air Boats, Inc., killed three passengers and the pilot, Charles Blair, who was also the President of Antilles Air Boats.**
>
> **The Grumman's left engine failed en route from St. Croix to St. Thomas**

and level flight could not be maintained with one engine. Blair then attempted to fly the airplane in ground effect, an aerodynamic technique that requires the airplane be flown close to the surface of the ground or water. Ground effect can reduce drag and make the lift of the wings more effective.

But the Board said in this case single engine flight could not have been maintained at any altitude because of the drag induced by the loss of the left engine cowl and the decreased efficiency of the right propeller which had not been properly maintained. Rather than attempt to fly the aircraft in ground effect, the pilot should have executed an emergency landing in the open sea. The airplane contacted the water with full power on the right engine and the left wing float struck the water causing the plane to cartwheel and breakup.

Captain Blair also failed to use good judgment in preparing his passengers for the emergency. Ample time was available for the Captain to instruct his passengers to don the life vests and make them aware of the locations of emergency exits but he took no action. The Captain again exhibited poor judgment when he did not prepare his passengers for the possibility the aircraft would strike the water.

> **Contributing to the accident were the company's inadequate maintenance program, the (Antilles Air Boats, Inc.) management influence which resulted in a disregard of FAA regulations, and FAA-approved maintenance policies, inadequate FAA surveillance of the airline, and deficient enforcement procedures.**
>
> **The Board said the FAA's surveillance and enforcement activities were inadequate. Recommended enforcement action was compromised regularly by FAA officials. Ultimately, the apparent policy of continual compromise of civil penalties rendered the FAA's enforcement process ineffective, and resulted in the recurrence of deficiencies in the Antilles Air Boats programs.**

During the 12 year period prior to the accident, I met Captain Blair on four occasions while investigating nonfatal occurrences involving Antilles Air Boats, Inc. airplanes. He was always friendly and cordial. He was one of the most exciting individuals I ever had the privilege of meeting. I still stand in awe of his historic accomplishments in aviation.

One of those nonfatal occurrences had an interesting twist. It involved a twin-engine amphibian that experienced an in-flight engine fire. The pilots executed a safe landing on protected waters at their destination. However, they were unable to water taxi the airplane on one engine. It just went round and round, always turning into the dead engine. Upon finding they could not extinguish the fire in a timely manner, the captain ordered an emergency evacuation. This forced the passengers to deplane into the waist-deep water and wade ashore. Normally, we did not respond to such occurrences; however, the incident garnered much news media attention, and I was dispatched so the Safety Board could reply: "We've got our investigator at the scene."

- -

FAULTY VISUAL CUES?

The pilot had 18,900 flying hours, including 3,500 in the reliable, twin-engine, Douglas DC-3 transport airplane he and his 22-year-old son were flying. The weather was clear with ten miles' visibililty and light surface winds at the time of the accident. The statement by the pilot follows:

> **Flight was planned in order to exercise aircraft systems and engines and also to maintain landing proficiency of pilot. Takeoff was normal. Aircraft remained in traffic pattern for touch-and-go landing. Traffic pattern was normal except downwind leg was extended a bit for landing traffic. Approach was slightly low but within limits of safety. Final approach (just before touchdown) was too low and I took action to bring aircraft up and applied power and back elevator control. Aircraft responded slightly and I increased control movement (back pressure). At this time the main gear struck soil lip of built-up runway and the airplane bounced upward. I realized that major damage had occurred to the main landing gear and the aircraft. Therefore, I took prompt action and crash landed on the hard sur-**

> **face runway. After the aircraft contacted the runway, I kept full back elevator control to prevent nosing over. Aircraft skidded approximately 1,500-2,000 feet.**
>
> **More and prompt attention should have been used and applied sooner on landing phase of flight. Pilot is well aware that corrective action was applied too late.**

Investigators are always grateful when they receive such forthright statements from pilots. When a pilot essentially "buys" an accident, it greatly influences the overall conduct of an investigation. In such instances, we normally accepted the pilots' statements carte blanche, and didn't go about trying to prove them wrong.

However, there was reason to doubt the pilot in this instance. In lieu of an accusing stance, I was always sympathetic towards pilots involved in accidents. That's why I preferred my role of investigator charged with collecting the facts, as opposed to that of an inspector charged, among other functions, with finding faults. My soft-soaped approach may have influenced the pilot to tell me his grave concerns.

The insurance policy for the airplane stipulated minimum pilot qualifications. His son, who had accumulated 350 flight hours, possessed a commercial pilot certificate with a rating in single-engine airplanes. Those qualifications did meet the minimum standards required by the insurance policy.

Since his son was qualified to be his copilot by FAA regulations, matters related to the small print in the insurance policy exceeded the scope of our investigation. So, without challenging him, I agreed he had reason for concerns that the insurance carrier might deny coverage. In view of what he told me, it was considered the pilot's son may have been flying the airplane when it crashed. However, such a circumstance would not relieve the pilot-in-command of his obligation to assure the safe conduct of the flight. If the copilot was flying the airplane, the Safety Board's probable cause determination would have cited the pilot-in-command's "inadequate supervision of the flight" as being involved in the accident.

GUILTY OF FORGETTING

The 63-year-old retired airline captain had accumulated 25,000 hours, including 1,000 hours in the amphibious airplane involved in the accident. He possessed type ratings in the Convair 880, 990; Boeing 707, 720, 757, 767; and the Lockheed L-1011. Prior to the accident, he made several water landings before proceeding to a nearby airport where he landed on a paved runway to obtain fuel. The pilot's account of the accident included the following:

> **I landed at the airport and refueled the aircraft. Took off to the west and made a right 90 degree turn while climbing to approximately 300 feet. I then started an approach for a landing on the lake in approximately one mile. I neglected to raise the landing gear for the water landing and upon touchdown, the aircraft capsized and floated inverted on the lake.**
>
> **In 41 years of flying I have never been involved in an accident or incident and have never received a violation of**

any kind. Pilots should use checklists for all operations regardless of their familiarity with the equipment.

Nothing further needs to be said. Sometimes, pilots land on runways with the landing gear retracted. At other times, pilots land on water surfaces with the gear down. The results are the same, busted airplanes and pilots who have learned they are not infallible.

ON BEING AN INSTRUCTOR PILOT

I gave considerable flight instruction while in the Marine Corps. In several instances after instruction flights in transport models, my crew chief, Master Sergeant Moose Marcellin, with whom I flew more than 2,000 hours, said: "Captain, you almost scared me to death." When I asked why, he said there were times he wondered why I hadn't taken the controls when he could see the flight was not progressing in a normal manner. That is a situation all instructor pilots face: just how far can they let their students go before safety is jeopardized.

A CASE IN POINT

The accident involved a twin-engine amphibian operated by one of the principal seaplane flight schools in our nation. The student, on his initial instruction flight in twin-engine seaplanes, said he completed the familiarization phase of instruction and practiced some normal landings on a lake. Thereafter, the instructor demonstrated water taxi techniques before the student took the controls and executed a "no-flaps" landing approach. He reported everything appeared normal when he retarded the throttles to idle in the landing flare. But before touching down, he heard a loud exclamation from the instructor pilot and found himself under water.

The student sustained serious injuries. He came to the surface after releasing his seatbelt and grabbed onto a flotation cushion from the wreckage. Shortly thereafter, he was rescued by personnel in a boat. They returned to the wreckage and one of the boat occupants went beneath the water surface and removed the fatally injured instructor pilot from his seat.

The 62-year-old pilot-in-command was a career flight instructor who had accumulated about 14,000 hours, including 8,400 hours as an instructor. He possessed the much coveted FAA authorization to act as a designated pilot examiner for applicants seeking single and multi-engine seaplane ratings. There were no witnesses to the occurrence. Impact damage showed the seaplane had abruptly nosed into the water.

Airplanes that give pilots ample warning of an approaching stall make excellent trainers. Along with other cues, pilots can sense in the seat of their pants an approaching stall. The stall characteristics of the airplane involved in the accident during a normal landing with the flaps extended, met this criterion. However, when the flaps were retracted, the stall features were far more abrupt. The investigation showed the instructor did not adequately supervise the flight during a critical phase of the "no-flaps" landing approach.

- -

TETHERED

The pilot landed his helicopter near the fuel pump. After attaching a grounding clip to the engine mount structure, he discovered the refueling hose would not reach his gas tank. He restarted the engine with the intention of air taxiing the skid-equipped helicopter a few feet closer. As he rose into a hover, the helicopter pitched down when the grounding line he had forgotten to disconnect became taut. Before the pilot realized what was happening, the helicopter rolled over on its side. Luckily, no one was injured by the flying debris when the helicopter tore itself apart while the rotor blades were thrashing about.

We could cite many other instances where well qualified pilots got caught with their guard down. However, these mishaps are sufficient to show a great degree of flight experience is not the only ingredient required to brew up an accident-free flying career. Which only confirms what we knew all along. No one is perfect.

ON BETTERING OUR WAYS

Achievements by the National Transportation Safety Board in promoting safety in the various transportation modes are primarily accomplished through the recommendation process. This is a grass-roots effort whereby anyone can propose a recommendation for consideration by the Board. When the Safety Board determines that a proposal addresses an existing problem area and suggests meaningful solutions, it is processed by its staff into a formal recommendation calling for changes in regulations, procedures, manufacturing processes, etc. The Safety Board forwards its recommendations to federal, state, and local agencies; airframe and component manufacturers; and other organizations or individuals in responsible positions, with the expectation that adoption of the proposals will result in improved safety. Most of the formal recommendations made by the Safety Board are derived from proposals generated in-house by its investigative staff.

Coming up with recommendation proposals was not a difficult task. After completing an investigation, the investigator has first-hand knowledge of all the relevant factors involved, and it's easy to spot where things might be done to prevent similar occurrences. And, since a recommendation proposal is just an in-house document until accepted by the Safety Board, it's possible to let your hair hang down from a literary standpoint, and use any writing style deemed appropriate. Included below is the crux of a safety recommendation proposal in which I used rather informal language regarding the ineffectiveness, from a safety standpoint, of the dry, stereotyped releases the Safety Board utilizes to publish its probable cause determinations for the vast majority of aviation accidents. Some editing, and the inclusion of clarifying remarks, has been utilized to adapt it for the general audience.

GENERAL RECOMMENDATION

In performing our tasks, we spend many hours feeding statistical data and accident information into our computers. Then about twelve times a year the computer is programmed to print out a book of accident releases in the brief format style that includes the Safety Board's probable cause determinations. And that's the way it goes. They have them, we record them, and regretfully, that's often the end of it. More often than not, the lessons to be learned from past miscues do not appear in the computerized releases and the odds are less than even that a similar mishap is already in the making. This paper proposes an avenue the Safety Board might traverse in its efforts to attract the attention of the aviation community by adding some "pizzazz" to its accident preventive efforts.

Those having a direct interest in a particular occurrence are anxiously awaiting the Safety Board's probable cause determination. While the brief format resumes provide the probable causes, they do nothing in the broad realm of accident prevention. Thumbing through an issue is somewhat like holding a wasp's nest. You don't know why you picked it up, and you're happy to drop it. Those releases just don't possess the spirited, vigorous qualities required to arouse the interest of pilots, aircrewmen, mechanics, and others engaged in the day-to-day activities associated with aircraft operations.

The pungent details of an accident don't appear in the brief format releases. Instead, the oftentimes exciting events that might have stimulated the interest of the aviation community with an infectious safety message go unnoticed and are forever buried in the Safety Board's files. In the meantime, pilots go about busting up their flying machines for the same old reasons, and investigations continue to show that the events that led to the dramatic and sudden climax had been present and operating over a long period of time.

Human error is the predominating factor in aircraft accidents and 80 percent is a good ballpark figure for those that are pilot-caused. Armed with this morsel of information, we don't need to speculate further as to the most fertile field for improvement in aviation safety. We already know the most common denominator. However, we can't make much headway with our preventive efforts until we figure out how to gusto up our accident releases so they will be more palatable to pilots. Using the old cliche, we need to figure out how to get the grease where the squeak is. And that's no easy task, because one of the surest ways to empty a pilot's lounge is to announce a safety meeting.

Aircraft accidents can be extremely interesting, as evidenced by the fact that several recent best sellers were written by authors who made liberal use of the facts, circumstances, and conditions contained in the Safety Board's reports for catastrophic air carrier occurrences. The same results can be achieved with many of the small aircraft accidents if we ferret out the salient facts and present them in an interesting manner. To protect the privacy of those involved, specific accidents, their locations, and the individuals involved, need not be identified. The articles should be sprinkled with colloquialisms and tart phrases. True, we won't win an award for literary achievement, but we'll get our message across because it will be enticing and acid clear.

This is not meant to imply accidents should be taken lightly. However, nothing can undo a mishap and the only reward is to profit from the miscues others have made. To capture the attention of the aviation community, the oftentimes tragic, ridiculous, bizarre, and sometimes humorous circumstances preceding accidents should be publicized in a manner designed to hold reader interest. This can be accomplished without showing a lack of compassion for those involved as exemplified in the U. S. Navy's APPROACH Magazine, The U. S. Air Force's AEROSPACE SAFETY Magazine, the Commonwealth of Australia's AVIATION SAFETY DIGEST, and other aviation periodicals.

The proposal suggested that the Safety Board publish a quarterly Safety Digest. In the event that was not feasible, it was suggested the Safety Board provide material to the Federal Aviation Administration for publication in its monthly General Aviation News Magazine, or to the publishers of other aviation magazines and periodicals.

Several short resumes were included that related to the circumstances involved in accidents considered interesting. Before reviewing those examples, it might be well to report how the recommendation proposal was viewed by personnel at the ivory tower in Washington. The Safety Board's Public Information Officer called and stated it was a good proposal and he had enjoyed reading it. He remarked that we were not funded or staffed to draft, assemble, and distribute such a publication. He further stated that a sufficient number of investigators did not possess the requisite writing skills to engage in the effort, and it would not be possible for his staff to accomplish the necessary editing. Additionally, he said the editors of aviation-oriented publications frowned upon accepting "canned" releases prepared by other sources, preferring instead to draft their own material from the facts, circumstances, and conditions of the accidents to which they were privy.

- -

THE STACKED DECK

The pilot made a refueling stop while on a visual flight in a light, single-engine airplane. He reported the vacuum pump, which powers most of the flight instruments, was inoperative. The operator at the airport was unable to repair or replace the pump within a time frame acceptable to the pilot. The preflight weather briefing showed instrument weather conditions prevailing over most of the planned route of flight due to low ceilings and visibility. The pilot discussed the possibility of between layers flight with the briefer, but was advised the layers would probably merge, and embedded thunderstorms were likely to be encountered. Thereafter, the pilot took off without filing a flight plan, telling the airport operator he would return if the weather was too severe.

The airplane crashed about 35 minutes after takeoff. Investigation showed the airplane sustained an in-flight structural breakup. Both wings and the entire tail separated and the two passengers in the rear seats were thrown out of the cabin before the fuselage impacted the ground. Radar weather observations showed several strong storm cells in the vicinity, and ground witnesses reported thunderstorm activity at the time of the crash. The spar fracture surfaces for the wings and tail were typical of those resulting from gross aerodynamic overloads, and no evidence of previous cracks or metal fatigue were found.

Essentially, the previous sentence, although possibly sounding like a governmental cover-up phrase, means no design deficiencies, manufacturing defects, or evidence that the structure had been subjected to previous loads in excess of the design limitations, were found.

By "Ivory soap" percentage probabilities, such in-flight breakups result from flight control inputs by the pilots as they attempt to regain control of their airplanes. In reality, the airplanes probably wouldn't have come apart if the pilots had just turned loose of the controls. However, this is not meant to infer the accidents wouldn't have happened. Instead, it only means, in many instances, the airplanes would have been intact when they crashed. In other words, an airplane normally won't come unglued without some help from the pilot.

The deck was stacked pretty heavily against the pilot in this instance. The overconfidence in his ability; his decision to initiate the flight with the known discrepancy in the power source for the primary flight instruments; and his decision to attempt the flight into forecasted instrument meteorological weather where thunderstorm activity was likely to be encountered, were the predominating factors in the accident.

Unlike the human body, vacuum pumps have no recuperative powers. The well-qualified, instrument-rated pilot knew he would be without his primary flight instruments. He failed to analyze properly his plight if thunderstorm activity was encountered. The tragedy was magnified because his poor judgment resulted in the death of his three passengers. If it were possible to undo all the mishaps resulting from pilots attempting operations with known deficiencies in their airplanes a vast improvement in safety would be realized.

- -

HAPPY HANK WITH THE EMPTY TANKS

The venerable, vintage, twin-engine, Douglas DC-3 airliner was on the seventh leg of an air-taxi flight with 30 passengers on board. It had not been refueled since the initial takeoff. The captain executed a wheels-up, forced landing when both engines failed because of fuel exhaustion (ran out of gas). There were no injuries to the flightcrew or passengers; however, the underside of the trusty old bird was severely rumpled.

There are absolutely no excuses for the captain's actions that jeopardized the lives of all on board. His wanton disregard for the FAA's stipulated regulations regarding minimum fuel requirements was the sole factor in the accident.

- -

OVERLOADED AND OUT OF BALANCE

Baggage and cargo were loaded on the light, twin-engine airplane before the pilot and three passengers boarded the flight. The airplane lifted off about halfway down the 6,200-foot runway. After becoming airborne, the nose began to rise slowly before pitching up abruptly. The airplane climbed to a height about 300 feet above the runway before falling off on the left wing and crashing. Everyone on board was killed.

Investigation showed the gross weight was 708 pounds over the authorized maximum, and the center of gravity was 3.9 inches aft of the rear limit. Knowledgeable aviation personnel would say those were a bunch of pounds and inches to be outside the weight and balance limitations in the relatively light airplane.

The tail heaviness caused by the unauthorized aft center of balance location resulted in the pilot not having enough forward elevator control to lower the nose after it pitched up. A lesson learned too late: that the weight and balance shall be maintained within the prescribed limits, or the earth shall rise up and smite thee. Accidents resulting from pilot failures to heed the prescribed gross weight and center of gravity limitations for their airplanes continue to occur, and oftentimes experienced pilots are involved.

- -

CONDENSATION PROBLEM

When patrons in bars order whiskey and water, they sometimes jokingly remark, "Water in moderation never hurt anyone." However, the same is certainly not true for airplanes, because just a little water in the fuel system is all that's required to make the silence deafening when the engine suddenly stops.

The pilot and three passengers made a short cross-country flight in a light, single-engine airplane. On the ensuing takeoff for return to the home base, the engine failed completely. The pilot attempted to turn back to the field, but the airplane stalled and nosed into the ground because he failed to maintain enough airspeed. The four occupants were killed. The carburetor bowl was found full of water. This evidence was sufficient to show that the engine failure was caused by water contamination of the fuel.

Prior to the initial takeoff, the airplane had been parked on the ramp for about two months with the fuel tanks one-half full. Water can enter the fuel system through leaky gas caps, condensation of water vapor out of the air in the tank, use of water-contaminated gasoline during refueling, etc. To inhibit water condensation, owners who follow good operating practices always keep the tanks full. Even this practice doesn't completely eliminate problems that can occur from water condensation in the fuel system.

Investigation showed the pilot accomplished a preflight inspection of the airplane before the initial takeoff. Water, being heavier than fuel, always settles to the bottom. During their preflight inspections, pilots are expected to utilize the drains that manufacturers provide at low points on the tanks and other fuel components to drain water out of the system. Pilots who don't ensure that the fuel system is free of water before taking off are just asking for trouble.

During accident investigations where all occupants are killed, it's not always possible to determine precisely the events that transpired. However, it was reasonable to assume that the pilot accomplished the first leg of the flight with the fuel selector positioned to a tank free of water. Thereafter, the second takeoff was made with the fuel selector positioned to another source contaminated by water that probably condensed out of the air in the partially filled tank.

COLD FACTS ABOUT A HOT ENGINE

The nearly new, light, twin-engine airplane was on an air-taxi flight in good visual weather. About 20 minutes after takeoff, the pilot shut the left engine down when it began to run rough. Shortly after turning back to the departure airport, the right engine began to malfunction. The left engine was restarted in an attempt to maintain level flight; however, the airplane still descended at a rate of 300 feet-per-minute. The pilot made a forced landing, but the airplane sustained damage when he intentionally swerved on the landing roll to avoid obstructions in his path. (In aviation terminology, it would be said he ground-looped the airplane.)

Investigation showed that the internal engine components had been subjected to extreme overheat distress. There were two "dead" cylinders (meaning no compression during crankshaft rotation) on the left engine and one "dead" cylinder on the right engine. The airplane had been flown only three hours since the last periodic inspection. During the inspection, maintenance personnel installed new spark plugs that were 30 to 40 percent "hotter" than those approved for use on the engine.

Engine manufacturers always stipulate which spark plugs are approved for use in their engines. Without going more deeply into the differences between hot and cold spark plugs, it's sufficient for these discussions to understand that the use of hotter plugs than called for in the specifications resulted in the excessively high operating temperatures that caused the engine malfunctions.

The mechanic's error shows that unless an engine is properly maintained, negligence, or a lack of necessary attention to detail, can steal its remaining life. In this instance, the proper spark plugs were in stock at the repair facility performing the inspection. The real puzzle was related to the mechanic's actions when he went to a nearby facility and purchased the incorrect plugs he installed in the engines.

AFRICAN ADVENTURE

There are many reasons why people go to Africa. Some go to see the Sphinx and pyramids, some for the thrill of shooting large animals, some are missionaries who want to spread the Good News, some in search of precious metals and gemstones, and others for multitudinous reasons. However, speaking for myself, comfortably situated in the Safety Board's Miami office working on the old golf handicap during leisure hours, there was only one reason to go to Africa. And that's because the powers that be sent me.

It didn't require much smarts to figure it wasn't going to be choice duty. That's because the Safety Board had folks assigned to another branch in its Washington headquarters who're supposed to handle such assignments. Those guys and gals keep the saddles strapped on their steeds for assignments in Europe, Japan, the South Pacific, or other exotic and romantic locations. But, let a commitment come up in some developing third-world nation, and they can come up with more reasons to weasel out than our elected officials can to justify another tax increase. However, all journeyman investigators are issued official passports because of the probability of such assignments. My dictionary defines a journeyman as a competent but undistinguished worker, which probably isn't too far off the mark.

I was travelling in the company of an FAA air carrier operations inspector. We made mistakes right away by letting our Washington staffs arrange our travel. They routed us via New York, and London. From London, we were scheduled to fly on the African nation's national airline to their capital city on the dark continent. The accident involved one of their Boeing 707 airplanes that crashed while landing at the capital city's international airport on a flight from London. My assignment was the

"Accredited Representative" from the State of Manufacture, and the FAA inspector's title was "Technical Specialist" under the rules and procedures of the International Civil Aviation Organization.

When the relatively long range 707 crashed, it played havoc with the air line's ability to service their international route structure, and we were delayed some two and a half days in London. We used the time to visit and sightsee in the land of tyrannous King George III, which we enjoyed very much. The natives exhibited a friendly attitude towards us "Yankees," and gave no outward display that they were discombobulated because they had gotten the short end of the wicket during the battles at Saratoga and Yorktown.

We were in London over a Sunday and had the good fortune of inadvertently getting caught in the Brighton Run. That's a 57-mile veteran car run from London to the coastal city of Brighton for vehicles predating 1905. The route wasn't closed to normal traffic, and our drive was time-consuming but extremely interesting. The impeccably restored vehicles included many three-wheeled models, and with their drivers and riders all dressed up for the occasion in the garb of those days, it was quite a sight to see. Some years later, I learned from a feature article in the National Geographic Magazine that the Brighton Run is the number-one choice of antique car buffs the world over. Those completing the run within eight hours win a coveted finisher's medal. Reportedly, the Brighton Run compares to Wimbledon, the Derby, and other noteworthy events on the English social calendar.

But back to work. The African nation's aviation authorities had requested assistance in the conduct of the investigation through our State Department. We were informed a representative from the Ambassador's staff would meet us. However, we committed another mistake when we failed to advise the Ambassador's office of our revised arrival time after the delay in London. Upon arrival at the airport serving the capital, no one met us and those official passports we were carrying didn't impress the nation's immigration personnel in any manner whatsoever.

A military government was running the country after a recent civil war. It appeared that customs personnel were confused because our passports weren't like those carried by other passengers. Telephone service was primitive and attempts to call our Ambassador's staff in the city were fruitless. It was an English-speaking nation, but it was apparent that their customs and immigration personnel were having difficulty with our versions of the language. When another of their flights arrived, we enlisted the flightcrew's assistance and were finally permitted to enter the country.

During our initial meeting with their aviation authorities, the designated Investigator-in-Charge was asked what assistance he desired. He responded that they wanted us to investigate the accident. When asked who would interrogate the crew and witnesses, he said we would. When asked who would write the report, he said he hoped I would. When asked who would sign the report, he said they would. In reality, there's room for lots of latitude in this arena. I asked similar questions during all my overseas assignments in an effort to render every possible degree of cooperation and assistance.

After the handshaking and preliminary discussions, we proceeded to the airport for an initial survey of the accident scene. There were no fatalities or serious injuries and although extensively damaged, the airplane was intact and probably repairable. The accident involved an overshoot of the runway while landing in a dense fog. The major damage to the airplane occurred as it wiped out a goodly portion of the approach light support structure after overrunning the end of the runway.

We met with the airport fire chief who had witnessed the accident. When asked where the airplane had touched down, he took us out on the runway and pointed out the spot, which was 6,700 feet down the 9,000-foot runway. My spontaneous response was, "They haven't made a 707 yet that you can stop in 2,300 feet; you'd probably need somewhere around 4,000 to 4,500 feet; however, the Boeing engineers will give us a more precise figure." One of their investigators said the captain reported the brake antiskid system wasn't working. But I replied, "No, it was working. We can see it cycling on and off by the skid marks on the pavement. Besides, the airplane's main wheel tires are still inflated, and they always blow when the pilot attempts to obtain maximum braking effectiveness with an inoperative antiskid system." They said the captain reported the engines wouldn't reverse. We responded that was a plus if they do reverse, but not a minus if they don't because the effects of reverse thrust may not be used when determining the Accelerate/Stop Distance (a computation required in the certification process for transport models by our FAA). However, we did suggest that they check the thrust reversers after the airplane was removed from the crash scene. From experience, we realized that back in the United States it takes time to collect the equipment required to move such a large airplane, and knew they wouldn't have this one out before we were back in our stateside offices.

The relatively sketchy evidence cited above provides the experienced investigator with enough information to make a preliminary decision as to the scope and magnitude of the required investigation. The evidence showed this was basically an operational accident, and the pertinent factors would be related to the prevailing weather, the weather data provided the flightcrew, and the reasons for the captain's decision to persist with the landing after overflying so much of the runway.

My judgment call regarding the scope of the investigation probably wouldn't have set too well with the hierachy of some of the unions that represent airline pilots. Since the preliminary data showed the probability of flight crewmember involvement, parties representing the pilots would have been looking for something or somebody else to blame the accident on if they adhered to their normal practices. However, speaking for myself and the FAA representative, in a faraway land where it wasn't safe to drink the water, it surely satisfied us. But more importantly, it satisfied the investigative team we were working with. Three days after arrival, we obtained accommodations at an American chain hotel where it was safe to drink the water. We had been spending much time outdoors and both of us had learned how difficult it is to quench a thirst with canned and bottled drinks.

The captain was a native and the copilot was a Frenchman. While I was relaxing at our first hotel, the copilot introduced himself and asked if he could discuss a matter that was causing him some concern. His request presented no problems and he related that he had been employed by the airline for some time and was about ready for promotion to captain. The airline was pretty heavily subsidized by the government, and he wanted to know if we could conduct our interview with him in a manner that would prevent him from having to analyze or second-guess the captain's actions.

The copilot said he had advised the captain several times during the approach that they wouldn't be able to stop on the runway, and all the conversation should be on the cockpit voice recorder. That is the "black box" that continuously records on magnetic tape the previous 30 minutes of radio communications with the flight as well as other conversation or noises in the cockpit. He was advised that it depended in a large part upon the evidence obtained during the interview of the captain. He was also reminded that we had no control over questions other parties to the investigation might ask.

The captain gave pretty straightforward answers during his interview. Basically, he stated he had abandoned two previous approaches because of the reduced visibility. The accident occurred on the final approach they could attempt before having to divert to their alternate airport because of the diminishing fuel supply. He couldn't believe they had actually landed so far down the runway.

So, as it turned out, we were able to acquiesce to the copilot's request to some extent without jeopardizing the overall conduct of the inquiry. From an investigative standpoint, the real pearls were unshucked several months later when the readout of the cockpit voice recorder at the Safety Board's laboratory in Washington showed the copilot giving the captain five separate and distinct warnings while on the final approach.

The captain did not order an emergency evacuation after the airplane came to a stop. In the United States, it's customary to order an emergency evacuation under such circumstances because of the possibility of fire from so much damage to the lower fuselage structure. It was noted during the examination of the wreckage that only the left forward cabin door had been used to evacuate the occupants.

While interviewing the cabin crewmembers, it was learned the flight attendant stationed by the second door on the left side was on her initial flight. She readily admitted she did not know how to operate the door. When asked if they hadn't covered that in her training, she replied she hadn't received any training. She testified she helped serve the food after the flight leveled off and did whatever else the senior flight attendants asked. Interrogation of the remaining flight attendants showed others were also deficient, but to lesser degrees, in their training requirements.

The cabin crewmembers on our airlines are flight attendants required by the FAA to be on board to ensure passenger safety throughout the normal conduct of the flight, and particularly in emergency situations. They must satisfactorily complete initial and recurrent training to ensure the high degree of knowledge and proficiency required to carry out their duties effectively. Most importantly, they are essential to the orderly emergency evacuation of the airplanes. They should never be considered stewards or stewardesses, although they do perform some related functions during routine operations. When we broke for lunch after interviewing the flight attendants, their Investigator-in-Charge was advised that such a breach in the stipulated training requirements would have provided the basis for our FAA to take rather stern actions against one of our air carriers.

About five days after our arrival, we had a fairly firm handle on the facts, circumstances, and conditions of the accident, and the FAA inspector returned home. My duties weren't completed, and I remained to study the air carrier's operations manual, collect other necessary data, and tie together the loose ends that would permit the writing of a report encompassing all the issues involved. We arranged for readout of the flight data recorder in London and transcription of the cockpit voice recorder in Washington.

The Investigator-in-Charge said he would advise me when the results were back, and I could return to complete the report. He was corrected immediately. I told him this excursion represented my two trips to Africa, namely the first and the last. An invitation was extended for all members of his investigative team to visit our office for the report-writing session. That seemed to strike a chord, and about two months later the entire four-man team arrived in Miami.

My return reservations on the African nation's airline were cancelled, and arrangements were made to fly home on a twice-weekly PAN AM flight. The problems experienced getting out of the country about equaled those getting in. I've never done a very good job of accounting for money, personally or otherwise, and couldn't satisfy their customs personnel. The official passport I was carrying did not impress the custom's agent. They even made me remove my shoes. Basically, the problems stemmed from the fact that I couldn't account for all the money brought into the country. Knowing I wouldn't be returning home via London, my British pounds had been used for tips. For awhile it appeared I would miss the flight, but the customs agent finally let me go. I was never in all my life so happy to get on an airplane. The entire African experience had been somewhat of a ordeal, and no attempt was made to hide my joy at being back on one of our airliners. The lovely flight attendant from Norway seemed to sense my feeling of relief, and a most pleasant flight was experienced back to New York.

I had been in the country ten days but really hadn't learned much about the natives or their customs and routines. My assigned driver was stopped by a soldier-policeman for some traffic infraction on one occasion, and that caused all manner of trouble. We were hauled off to the local constabulary and weren't released until an authority from the aviation department arrived. It was learned some of the problems stemmed from the fact that the soldier didn't like my attitude. I have never been prone to act in an officious manner, and from my point of view, the problem was related to our difficulty in understanding each other's rendition of the English language. At any rate, I speculate there will be a few more revolutionary wars before our military personnel are ever permitted to be so much in presence with such ultimate authority over the general populace.

While in the country, my efforts had been directed towards completing the on-scene investigation, and when that was accomplished, I was eager to go home. A few dozen oil drilling experts, construction workers, and technicians had boarded the flight for stateside visits with their families. They were a robust group, fully enjoying themselves. The term "good old boys'" would be a most apt description. Upon disclosing my troubles getting out of the country, they asked why I hadn't used an expediter, explaining these were locals who could grease the skids so you could almost slide right out of the terminal. Apparently, the expediters split their fees with the customs and immigration personnel, but for me, that was another lesson learned too late. The "good old boys'" response to my request for a description of the native population was amusing. They said the native inhabitants woke up every morning with a clean slate. It didn't make any difference what they did yesterday, last week, or last month; the natives started every day right back on square one.

After collecting and sorting all the evidence, it's easy for the journeyman investigator to determine what's pertinent and what's not. Thereafter, writing the investigative report, analysis, and conclusions is not a difficult assignment. Almost all nations, including the United States, use the format devised by the International Civil Aviation Organization. There may have been some coverup of the airline's operational procedures and training deficiencies. However, the report addressed the major issues, and as we would say in the trade, it would probably "float."

The Investigator-in-Charge and other team members seemed to thoroughly enjoy their visit to Miami and we loaded them up with technical data and manuals that might aid them in the future. Looking back with the realization I'll never have to go back again, it was a fairly interesting trip.

It's not necessary to reminisce about everything learned from this African adventure. Most of it's fairly obvious from reading between the lines. However, the issue of the airline companies we elect to patronize merits discussion. It's just inherent for all nations, and especially the emerging third-world countries, to prefer that native flightcrews man their airliners. We must always bear in mind that, insofar as the flightcrews are concerned, all we're ever guaranteed in the United States or elsewhere is that they possess or demonstrate only the minimum knowledge or degree of skill required to meet the established criteria.

Just pondering that precept should create some cause for concern. The problem facing passengers is that they have no way of determining who's unqualified, minimally qualified, or fully qualified. I personally believe there's reason to question the qualifications of some individual flightcrew members on a worldwide basis. Therefore, since the passenger has no way of determining who's qualified and who's not, it would appear prudent to select those airline companies that have established and maintained a proven safety record over the long haul.

THOSE DULL PUBLIC HEARINGS

On occasion, the powers that be from the Safety Board's Washington headquarters would visit the various regional offices. During such visits, it wasn't unusual for a general meeting to be held to chitchat about ongoing activities and policies of the Board. Dick Spears made a visit to the Miami, Florida office after being appointed Managing Director by Chairman Todd. Dick exhibited a good-natured outlook while showing a keen interest in our office activities and the problems we were experiencing.

At the conclusion of the meeting, Dick asked each of us what actions we would take to improve the Board if we were Chairman. When it came my turn, I said revisions would be made to permit the Board to pay in a more timely manner the obligations encumbered by its investigators. After some discussion, during which the other investigators agreed it posed a problem, Dick said he was in a position to take corrective actions, and after returning to Washington, he did. He then asked if there was anything else I would do. I replied: "If I was Chairman of the Board, we would never have another one of those dull public hearings." And to say that failed to strike a harmonious chord would be putting it mildly. He advised me I would never go far in Washington with such an attitude.

Dick asked what I found objectionable about the Safety Board's public hearings. He was told we had just attended some sessions of the public hearing conducted as part of the Safety Board's investigation of the Eastern Air Lines L-1011 tri-jet accident that occurred in the Everglades Swamp near Miami, Florida. This was the accident where the flightcrew, while attempting to cope with a faulty nose landing gear indicator light, inadvertently let the airplane descend into the swamp. He was advised that absolutely nothing new was learned during attendance. However, even more disturbing was the fact that some of the parties to the hearing appeared to take an adversative stance towards the proceedings. It was pointed out that such an attitude is contrary to, and in direct conflict with, that which must prevail among the various parties during the free and open discussions of the relevant factors that is an essential part of every investigation.

While maybe not losing any Brownie points, the discussions did move on to another member of our office team. Incidentally, my role in the Lockheed L-1011 accident investigation was the "Gopher," signifying, since the accident occurred in our locale, that I should know where to go-for-this or go-for-that to provide logistics support and necessary liaison with local authorities for the various working groups conducting the investigation. However, my actual title was Assistant Investigator-in-Charge, and I had immediate access to all the evidence relating to the accident.

Safety Board regulations state: "The Board may order a public hearing as part of an accident investigation whenever such hearing is deemed necessary in the public interest." The Safety Board normally orders a public hearing as part of its investigation of catastrophic transportation accidents in any mode. Additionally, the Board may order a public hearing for "noncatastrophic" occurrences that draw wide public interest, or for other accidents that address significant safety issues. The Board's description of the nature and purpose of a public hearing is as follows:

> **Transportation accident hearings are convened to assist the Board in determining cause or probable cause of an accident, in reporting the facts, circum-**

stances, and conditions, of the accident and in ascertaining measures which will tend to prevent accidents and promote transportation safety. Such hearings are factfinding proceedings with no formal issues and no adverse parties.

So now we know what a public hearing is supposed to be and what is hoped will be accomplished. However, that is not the real world. In reality, a public hearing is nothing but an ostentatious display by the Safety Board wherein the wheel is reinvented insofar as obtaining evidence relating to an accident is concerned. In effect, it's an instant replay where all the facts, circumstances, and conditions, already collected and known to the Safety Board and other parties to an investigation, are recollected in a more or less quasi-legal session presided over by a Safety Board member.

New evidence is rarely uncovered during a public hearing. From an investigator's viewpoint, it appears the primary purpose of the proceedings is to direct news media attention to the Safety Board and its activities. Therefore, in those instances where it might be feasible to hold a public hearing in more than one location, the issue is usually settled in favor of the location that will give the Safety Board the most publicity. As apparent from the title, the general public is permitted to attend, but not participate in, the activities. The actual proceedings constitute a poor showing insofar as being an interesting spectator sport. That's because there is just nothing very exciting about the nitty-gritty details of any accident investigation except for those in the audience who might have some vested interest in the outcome.

Holding a public hearing is no small matter. Much organization and planning are necessary. A Safety Board member is appointed chairman of the inquiry; a hearing officer from the Board's Washington D.C. staff is designated; a member of the General Counsel's office for the Board is usually involved; a technical panel consisting of Board investigators and specialists is appointed to initially develop the testimony of witnesses; parties consisting of those persons, agencies, companies, and associations whose participation is deemed necessary are designated; and subpoenas for the attendance of witnesses and the production of documents must be issued. Additionally, a prehearing conference must be held, a suitable location where the public can be accommodated must be reserved, travel and lodging arrangements must be made, etc.

When all these requirements are considered, a public hearing evolves into a very costly affair insofar as the expenditure of both manpower and monetary resources are concerned for both the Safety Board and the designated parties to the hearing. All of this may make us wonder: Are all these carryings-on necessary just to rehash evidence that has already been uncovered?

The foregoing raises a question as to what a public hearing really is. Well, in my opinion, a public hearing is nothing but a Dog and Pony Show. This raises the question, what's a Dog and Pony Show? Why, that's any ridiculous and costly extravaganza serving no useful purpose other than directing attention to those who scheduled it.

JUST THE FACTS, MA'AM

It was midafternoon on Friday the 13th and we were winding down after an investigation in a distant city. Before returning home, we wanted to interview the pilot. I visited the hospital alone so the other parties to the investigation could catch earlier flights home. Upon arrival, it was learned the pilot was on a ward where visitors were permitted. I poked my head in his door, identified myself, and he invited me in. I showed him my credentials, and asked if he felt up to discussing the accident. He gave an affirmative reply; however, before we got through the preliminaries, all Hades broke loose. No offense intended, Florence, but I was apprehended by the Head Nurse. She summarily evicted me from the patient's room while administering a very severe tongue lashing for my failure to obtain prior permission from the medical staff.

Collecting witness statements can be a very time-consuming aspect of an investigation. Learning how to size up witnesses effectively is a near art form in itself which I never fully mastered. During my investigative career, I met several FAA inspectors who were particularly adept at obtaining meaningful information from witnesses. When one of them was assigned, I requested that the FAA's representative collect the witness statements while we, meaning the other parties to the investigation and myself, conducted the wreckage examination.

Some witnesses can be handed a statement form and their accounts will be just what we were hoping to obtain. In other instances, it's oftentimes best to conduct a verbal interview and then write a substance account of a witness' observations for their signature. Children from about age six up make excellent witnesses. Even better than many adults, they tend to stick to the facts without fantasizing or analyzing. Instances were observed where they steadfastly refused to alter their accounts as suggested by their parents, who were only attempting to ensure that they accurately described their observations.

Many witnesses have a tendency to state the airplane crashed, and then go into the details of their efforts to rescue or assist the victims. While such actions are commendable, the investigator is usually attempting to sort out the pertinent events and circumstances that preceded the crash. On a few occasions, witnesses were encountered who, while not exactly hostile, exhibited uncooperative attitudes.

One investigation involved a four-place, single-engine airplane that was taking off from a relatively short turf airstrip on a business flight. The pilot was the owner of an engineering firm, and his two passengers were employees. All three occupants were hospitalized after the airplane crashed in a wooded area adjacent to the right side of the airstrip.

An analysis of statements by ground witnesses indicated the pilot may have executed a premature liftoff before gaining flying speed. The pilot and front seat passenger could not be interviewed because of the seriousness of their injuries; however, medical authorities granted us permission to interview the other passenger.

From the very outset of our interrogation, it was apparent he was reluctant to discuss the details of the crash. He had no aviation background, and we considered his reticence was occasioned by an attempt to say nothing that would implicate his employer in the cause of the accident. We were primarily concerned about the operation of the engine; but when we asked whether he had noticed anything unusual, he replied he didn't know, saying he had been studying a blueprint.

While walking out of the hospital, the FAA inspector and I concluded he had given us all the information we needed. That's because human nature will not allow us to escape our inherent self-preservation instincts. If the engine had sputtered or otherwise malfunctioned in any manner whatsoever during the takeoff attempt, it would have corralled his immediate and wholehearted attention. Our conclusions were substantiated when, after recovering, the pilot stated he did not experience any difficulties with the operation of the engine prior to the crash.

- -

Nurses in general and head nurses in particular represent formidable adversaries, especially in those instances when we intruded in their domain. I bear no grudges; they were only carrying out their duties. After gaining investigative experience, and instituting proper protocol before interviewing patients, I discovered just how helpful and cooperative nurses could be.

As for doctors, well, they represented a different breed altogether. There were occasions when the assistance they rendered exceeded anything that could be expected. In instances where the doctors were pilots, or keenly interested in an accident for other reasons, we sometimes became engrossed in lengthy discussions of the mishap under investigation, different accidents the doctors were interested in, or other aviation topics they asked about. During such discussions, I withheld nothing, and left the subject matter and length of our sessions up to the doctors.

Before I returned home after one investigation, a doctor was contacted to see whether the pilot was well enough to answer several questions. The doctor said the patient was in the intensive care unit and he wouldn't know until he made his rounds. We arranged a meeting in the waiting room and visited the patient together. We found the pilot orientated as to his surroundings. My interview was conducted from one side of the patient's bed while the doctor checked his vital signs and examined his wounds from the other. However, there were many instances when doctors advised that their patients were too seriously injured, or too heavily sedated, to give rational answers.

Getting into a patient's room in a military hospital on weekends or other times when the medical and administrative staffs are not available can prove to be a formidable task. Apparently, military hospital rules are set in concrete. It's either black or white; yes or no; there is no grey area, or remote possibility of a "maybe". One Saturday, I was sitting in a hospital lobby pondering my next move after being refused permission by the receptionist to visit a patient who had been injured in an accident while flying his privately owned airplane. Things were relatively quiet, with very little coming and going. An elegant lady about the pilot's age came down on the elevator and entered the vending room. I followed her, and in answer to my query, she told me she was the pilot's wife. After I introduced myself and showed her my credentials, she gave me much needed information, said her husband very much wanted to discuss the accident with me, and made arrangements for our meeting.

- -

IMPROPER LOOKOUT

Flying has been characterized by some as hours and hours of boredom punctuated by moments of stark terror. While not ascribing to such a definition, I will admit flying can become mighty exciting if the pilot doesn't stay on his toes and mind the store. When the pilot's attention is diverted for whatever reason from flying the airplane, the situation can become precarious.

One accident involved a helicopter that abruptly descended into ocean waters while flying just offshore about 50 feet above the water surface. The left rear door had been removed to assist the passenger occupying the seat behind the pilot in photographing beach erosion. At the time of the crash, witnesses observed the pilot turned aft, with his left arm behind the cockpit seats. The pilot was killed and the two passengers sustained serious injuries. After impacting the water, both passengers came to the surface without knowing how they escaped from the wreckage.

The violent carryings-on that occur when a helicopter's rotor blades strike the surface during a crash have been likened to the flailing about of a frustrated palm tree in a hurricane. During the examination of the wreckage of the two-bladed helicopter, we found an irregularity in the pitch chain linkage to the main rotor blades. It involved a small, screw-in component that is safety-wired to prevent further movement after mechanics make the final adjustments.

The linkage to one of the blades broke in two during the crash sequence. On the other side, only the safety wire was broken and the threads had backed completely off. The component showed no visual distortions, and it was confirmed during subsequent metallurgical examinations that it was undamaged. This is one of those things an experienced investigator will not disregard. He instantly recognizes it as a topic he will likely be grilled about by those ambulance-chasing liability lawyers.

Representatives from the FAA and the manufacturer were participating in the investigation. Although difficult to comprehend, we knew the undamaged linkage had somehow managed to back completely out during the crash sequence. Our analyses were based on knowledge that the lengths of the pitch chain linkage to the two blades are very critical. Severe vibrations would have occurred if the linkage to one blade began backing off in flight, with the probability that the helicopter would have become uncontrollable before final separation occurred. After using the screw-in linkage to make the final adjustments to the main rotor blades, mechanics are required to determine that the two adjustable screw-in components, one for each blade, are of nearly equal length before safety wire is used to prevent further movement. However, we were still very much concerned because finding the separated, but otherwise undamaged, linkage component defied logic.

With hopes that we might resolve the matter, I visited the hospital to see if we could get a simple "yes" or "no" answer from one of the passengers as to whether they noticed any unusual vibrations before the crash. The front seat passenger was unable to answer any questions and I wasn't permitted to interview the other. However, after I conferred with his doctor and explained what we wanted to know, he came back and said the patient gave a "no" reply. Several months later I returned to interview the front seat passenger, who was well on the road to recovery. The substance of his observations was as follows:

> **There were no unusual vibrations. He observed the pilot turned around in his seat with his left arm back as he attempted to stow a cushion that had gotten loose. Upon looking forward again, the passenger realized that the helicopter was descending into the water and screamed at the top of his lungs. He related the pitch down of the nose must have been gradual, because he did not feel any motion to that effect.**

The passengers' observations confirmed our beliefs that the separated, but otherwise undamaged, pitch change linkage component was not involved in the cause of the crash. The accident was attributed to inadequate lookout by the pilot, who inadvertently descended into the water while attempting to stow the loose cushion.

NO HEMMING AND HAWING ABOUT

The instructor in a single engine trainer had demonstrated a touch-and-go landing to the new student pilot. Witnesses observed the airplane in a very nose-high attitude on the climbout with the engine sounding as though it was operating at full power. The airplane suddenly stalled, nosed over, and spun into the ground. The airport was uncontrolled (not equipped with a control tower), and, at the time of the crash, another airplane was in the pattern executing a landing approach in the opposite direction.

One witness used what could be called conventional language in his account of the accident. However, another witness was far more candid. So much so, some cleanup of the language has been accomplished in the below listed version:

> **The dumb flight instructor did a downwind takeoff against landing traffic. He had to turn left towards rising terrain and, with the engine screaming at full power, the airplane attained a banked, nose high attitude. Followed shortly thereafter by the wings waggling, then the old wingover nose down to a one turn try to screw itself into the ground. Thereafter, the airplane did a typical end over end tea kettle maneuver. The brand new student pilot got one heck of an "E" coupon ride.**

By way of explanation, the "E" coupon made reference to the tickets required for the most thrilling rides at the Disneyland amusement park in California.

There was one other aspect of this investigation meriting discussion. In an obvious attempt to avoid liability, neither the owner nor the instructor pilot wanted to be designated the "operator" of the airplane. According to the Safety Board, "operator" means any person who causes or authorizes the operation of an aircraft, such as the owner, lessee, or bailee of an aircraft. In this instance, the aircraft owner stated he rented the airplane to the instructor pilot with no knowledge that it would be a dual training flight. The instructor pilot claimed he was flying the airplane on the dual instruction flight for the aircraft owner. While we always attempted to determine the operator, it is neither the investigator's nor the Safety Board's responsibility to resolve the matter. In my factual report, I alluded to the controversy without attempting to provide an answer; thereby leaving settlement of the dispute to the parties involved.

LIKE AN OAK LEAF IN A TWISTER

During the final stages of a visual landing approach behind a Boeing 727 jet transport aircraft, a small airplane was suddenly flipped inverted before crashing beside the runway. While vectoring the airplane to land behind the air carrier flight, the approach controller advised the pilot he could land on the parallel runway if he desired. However, the pilot elected to continue with the approach and land behind the jet. After contacting the tower controller, the pilot was cleared to land and advised: "Caution Wake Turbulence."

The passenger on the small airplane stated they were just short of the runway when the pilot turned to her and said, "Feel that, it's from the jet." She said she acknowledged some turbulence, being especially aware of it because the morning flight had been so smooth and turbulence-free. She related they were about 50 feet above the runway with the wing flaps full down and the power off when they were suddenly caught in the turbulence and flipped over.

The "Invisible Force" had claimed another victim, the "Invisible Force" in this instance being the wake turbulence produced by the jet that landed ahead of the accident airplane. It should not be confused with prop wash from propeller-driven airplanes or the blast from turbojet engines. Wake turbulence is a by-product of the lift produced by the wings.

The lifting force of the wings is created by a pressure differential over the wing surface. When generating lift, a lower air pressure exists on the top surface of the wing than on the underside. This pressure differential triggers a roll up of the airflow aft of the wing tips. After the roll up is completed, the wake consists of two swirling, counter rotating, cylindrical vortices that initially sink at a rate of about 400 to 500 feet per minute with their strength diminishing with time. The phenomenon is depicted in the following sketch:

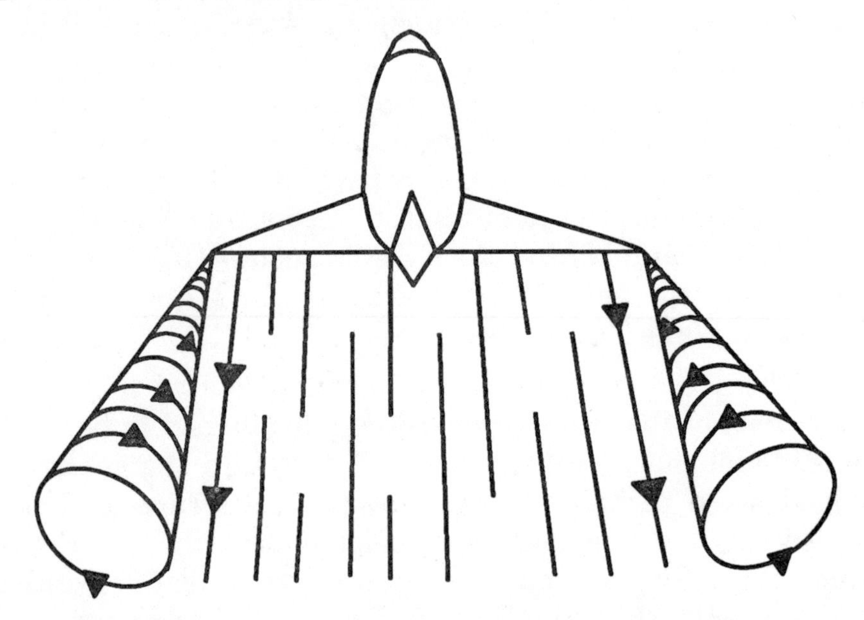

The strength of the vortices is governed by the weight, speed, and shape of the wing of the generating aircraft. Deployment of the wing trailing edge flaps and leading edge slats will alter the shape of a wing. However, the primary factor is the aircraft's gross weight. As the weight of the generating aircraft goes up, the strength of the vortices increases. While all airplanes generate wake turbulence vortices, those produced by light airplanes are of insufficient strength to present a significant hazard to following aircraft. The greatest vortex strength occurs when the generating aircraft is HEAVY, SLOW, and CLEAN (meaning with the wing flaps, slats, and landing gear retracted).

Wake turbulence generated by air carrier jet transports can be hazardous to light airplanes. Other jet transport models are also in jeopardy, especially when the airplanes generating the vortices are of the heavy variety like the Boeing 747, McDonnell-Douglas DC-10, Lockheed L-1011, etc. A wake turbulence encounter can cause catastrophic structural failure or, as stated in the title of this chapter, toss a small airplane about **"LIKE AN OAK LEAF IN A TWISTER."**

The statement by the passenger on the accident airplane included the following: "I feel the air traffic controller made a crucial error in her decision to land a small aircraft behind the jet within minimal legal distance allowed between aircraft. It was an apparent afterthought that she gave the pilot the option of the clear (parallel) runway after we were so near to landing. I feel that she, in her position, was aware of the susceptibility of a small aircraft landing behind a jet. The accident may have been avoided if the pilot had landed on the other clear runway."

Only part of what's stated in the above paragraph is correct. The pilot was executing the approach in good visual weather conditions, and the tower controller fulfilled her duties and responsibilities when she radioed, "Caution Wake Turbulence." The Airman's Information Manual states:

> **WHETHER OR NOT A WARNING HAS BEEN GIVEN, THE PILOT IS EXPECTED TO ADJUST HIS OPERATIONS AND FLIGHT PATH AS NECESSARY TO PRECLUDE SERIOUS WAKE TURBULENCE ENCOUNTERS.**

While it's true the accident probably wouldn't have happened if the pilot had landed on the parallel runway, there was nothing improper about the clearance the flight received to land behind the jet. Pilots are expected to follow the procedures for wake turbulence avoidance that are described in the FAA's Airman's Information Manual and Advisory Circulars. It is incumbent upon pilots to understand that to avoid wake turbulence, they should remain above the flight paths of preceding jets and land beyond the point on the runways where the jets touch down.

The pilot reported 1,200 flight hours. That's quite a bit of flight experience, certainly enough to take him out of the novice category. He showed he knew something about wake turbulence when he said: "Feel that, it's from the jet." Maybe that old Nemesis "Complacency" caught up with him. My dictionary shows the Greek Mythology definition for Nemesis to be: "The Goddess of retributive justice or vengeance."

At any rate, there are many general aviation pilots who expend hardly any effort to become knowledgeable, or remain current, about a wide range of aviation subjects vital to their safety. It's probably true that pilots who don't personally subscribe to the Airman's Information Manual and Advisory Circulars have improper attitudes towards their flying, unless, of course, they otherwise have access to those publications.

Wake turbulence caused a fatal accident involving a twin-engine, propeller-driven airplane that crashed after the pilot experienced a loss of control. Witnesses observed the accident airplane suddenly begin rocking violently about while crossing behind and beneath the flight path of a Lockheed L-1011 tri-jet. This is the precise location where wake turbulence is most likely to be encountered. The pilot failed to regain control of the airplane before it crashed.

In another accident, witnesses observed an airplane dive into the ground in a near vertical attitude in clear weather. The crash site was directly beneath the approach path for air carrier transport airplanes landing at a nearby international airport. The occupants were killed and the wreckage was subjected to an intense postcrash fire. Insufficient evidence was uncovered to determine precisely the factors relating to the loss of control by the pilot. It was considered possible that an encounter with the wake turbulence produced by one of the commercial jets landing at the international airport was involved.

The personal effects of the occupants were destroyed in the post crash fire. We experienced a great deal of difficulty determining the identity of both the airplane and its occupants. During the initial phase of most investigations, visits by, or inquiries from, those having knowledge of the airplane or its occupants are normal. In this instance, the airplane was not on a flight plan and we had no knowledge whatsoever as to the nature of the flight.

A steel data plate is attached to both the airplane and the engine. We were unable to find the data plate that would have identified the airplane. The engine data plate was found after it was removed from the impact crater. However, it did not necessarily identify the airplane, because engines are subject to removal from one airplane and installation on another.

It took several days, but we were ultimately able to identify the airplane by reference to the engine manufacturer's records. We then traced the airplane to its home base in California, where it was learned the two occupants were visiting the pilot's parents who lived in Florida. A telephone call confirmed that the occupants had departed for the return flight on the accident date. The pilot's father said they heard the initial news accounts of the accident. However, since three people were reported to be on board, they hadn't become alarmed. The accident represented one of about a half dozen instances where it was necessary for me to report fatal injuries to the victim's next of kin, a responsibility I would very much prefer to leave to those of the cloth.

DETECTIVE STORIES

My brother, a Presbyterian minister, resided in a city in our area of responsibility. Whenever conducting an investigation in his locale, I tried to visit him. Invariably, our conversation would drift into my field of endeavor. I once mentioned having a live pilot was not always beneficial to an investigation. The statement was based on past experiences where it had been determined that pilots hadn't stuck to the truth when reporting the circumstances leading up to an accident. On this particular occasion, the investigation extended into the weekend and I attended the Sunday morning worship service. It was surprising, and I was somewhat embarrassed, to learn he had incorporated my statement regarding the veracity of a few pilots into his morning message.

When it becomes apparent a pilot's account is not accurate, the investigator must deviate from his usual role as a collector of factual information, don his detective cap, and attempt to ferret out the truth. It was my practice to let pilots report the facts, circumstances, and conditions of accidents without interruption, although there were occasions when it was necessary to ask for clarification of some particular point. Never, on any occasion, was a pilot or other flightcrew member challenged or intimidated regarding anything stated during the initial interview. This took some self control because, at times, it was readily apparent to the experienced investigator that the facts reported were not correct, or the events could not have happened as stated. No attempt is being made to imply straying from the truth was the general rule. However, journeyman investigators have learned to remain alert while delving deeply enough into the circumstances of an accident to assure that the events happened as reported.

On one occasion, notification was received in the middle of the night that a pilot had experienced a collapse of the landing gear upon arrival at an airport in the Florida panhandle. There were no injuries or fatalities, but an on-scene investigation was required because it was an air-taxi flight. An early morning commercial flight, one we called the 0-Dark-30 red-eye special, was boarded, and I arrived at the scene about 8:30 the next morning. Permission had been granted to move the wreckage because it was blocking an active runway and the airplane was resting on makeshift jacks on the parking ramp. The pilot was waiting at the FAA's flight service station on the airport and gave a very detailed account of the collapse of the landing gear while on the landing roll. However, only a cursory examination of the wreckage was required to ascertain the main landing gear had sustained a gross overload failure in the rearward direction as opposed to a less damaging collapse on the landing roll.

The pilot was apprised of the finding and again was asked where he initially touched down. He maintained he had landed on the runway pavement. Very little debris was found on the runway, and we suspected the airplane may have touched down short of the runway threshold. Arrangements were made with airport authorities and control tower personnel to search the approach area short of the runway for possible clues. The pilot's request to go along was denied, and it took only a short while to find tire tracks and some landing gear debris in the soft terrain about 400 feet short of the runway threshold.

What actually happened was the pilot, upon discovering he had not landed on the runway pavement, reapplied power and became airborne again. However, the landing gear had received major damage on the initial touchdown. Thereafter, when he landed on the runway, the airplane slid to a stop on its belly. When confronted with the evidence we had uncovered, the pilot took advantage of the opportunity afforded to revise his statement and the on-scene phase of the investigation was terminated.

A CASE OF WANTON DISREGARD FOR PASSENGER SAFETY

There was a tragic accident in which five passengers perished. The ten-place, twin-engine, air-taxi airplane departed from St. Croix in the U. S. Virgin Islands on a morning flight to Puerto Rico with the pilot and nine passengers on board. The pilot was forced to ditch the airplane about 15 minutes after takeoff when both engines failed almost simultaneously. Adequate flotation equipment was available for all the occupants. However, five passengers failed to escape. The airplane sank in deep ocean waters and was not recovered. From the very outset of the investigation, we considered the accident was related to either fuel exhaustion (ran out of fuel), or mismanagement of the fuel system by the pilot.

Some explanation of the rationale behind our belief that fuel problems were involved might be appropriate. The engines, and the fuel supply for each engine on light, twin-engine airplanes, are almost totally isolated from each other. Unlike some multi-engine models, the airplane involved in the accident was not equipped with a fuselage fuel tank capable of providing fuel to both engines, either singularly or simultaneously. All the fuel was stored in wing tanks. In such configurations, it is normal to supply fuel to the left engine from the left wing tanks, and feed fuel to the right engine from the right wing tanks. Provisions are made to crossfeed fuel from the left wing tanks to the right engine, and vice versa. However, the crossfeed feature is installed for emergency operations, or occasions when for some reason there are unequal quantities of fuel in the wing tanks.

Additionally, all small reciprocating aircraft engines are equipped with two magnetos each. The magnetos, which provide electrical current for spark plug ignition, operate off the engine accessory drives and are totally independent of the airplane's normal electrical system. The failure of one magneto occurs occasionally; however, this results in only a slight decrease in power. Therefore, the chances of both engines sustaining a complete loss of power simultaneously for reasons other than fuel starvation, or fuel system mismanagement by the pilot, were considered to be remote. These theories don't always pan out and investigators sometimes find themselves barking up the wrong tree. The trick is to keep an open mind and ensure that pertinent evidence is not lost, destroyed, or overlooked, while searching for the truth.

The pilot and operator reported the airplane had been used on four round-trip, air-taxi flights since the fuel tanks were topped off. Three of the flights were to other locations in Puerto Rico, and the other to St. Croix. Knowing the fuel burnoff per hour, it was simple to compute that the reported flights would not have been of sufficient duration to exhaust the fuel supply. Intensive study of the operator's records and a search for witnesses who could provide pertinent information were fruitless. Likewise, no evidence of another flight was found during the time-consuming review of tape-recorded communications at Puerto Rican airport control towers.

We were aware that airplanes had to clear through U. S. Customs before departing from St. Croix on flights to Puerto Rico. A review of records on file at the U. S. Customs office in St. Croix showed the airplane made another round-trip flight between Puerto Rico and St. Croix that was unreported by both the pilot and operator. Armed with this information, we had "prima facie" evidence that the engine failures were related to fuel exhaustion. There was just no way the airplane's total fuel supply would be sufficient for all the flights made since the last refueling. Thereafter, the detective phase of the investigation was terminated, and we continued collecting the necessary data for the comprehensive report required for such tragic occurrences.

The gross neglect exhibited by the pilot and operator showed a wanton disregard for safety. Since the same pilot made all flights since the last refueling, he had to know he was flirting with disaster. FAA regulations relating to fuel requirements are quite specific, and the pilot's casual approach to his responsibilities was inexcusable. Causing the death of five innocent passengers under such circumstances was almost criminal.

It is not the Safety Board's role to recommend or initiate any actions against those who cause accidents. The pilot or operator may be cited by the FAA for violations of the regulations; however, such proceedings usually result in only temporary suspension of their licenses or certificates. The processing of any criminal proceedings is the responsibility of local law enforcement agencies, and they rarely become involved.

Insurance investigators and adjustors are not permitted to become parties to Safety Board investigations. However, the Investigator-in-Charge routinely conducts liaison with insurance adjustors concerning preservation of the wreckage and other matters of mutual concern. Nothing prevents insurance personnel from conducting independent investigations, or looking into other matters, providing such activities don't conflict or interfere with the Safety Board's investigations. On some occasions, it was noted that insurance adjustors attempted to make prompt settlements with the victim's next-of-kin. Obviously, they were intent on obtaining releases from further liability for their insurance companies. While I don't consider that I'm qualified to give legal advice, it might have been prudent for the aggrieved to delay liability settlements until the Safety Board's investigations were completed, and the pertinent facts, circumstances, and conditions of the accidents were known.

HIGH FLYING INACCURACIES

Let us now explore the high-water mark in the art of fibbery. Or, as the well-known movie personality might have said: "Baby, you ain't heard nothin' in the way of tall tales until you've listened to those drug smuggling pilots relate the circumstances leading up to accidents." For that reason, it doesn't take another Sherlock Holmes, Sam Spade, Lieutenant Columbo, or other top-notch detective to recognize when drug-hauling pilots are telling nothing but baldfaced lies. Those airmen are so accustomed to falsifying the record, the reverberations that occur when they inadvertently tell the truth have resulted in readings as high as 3.7 on seismograph Richter scales. Or, as seen in another light, lie detector machines strapped onto drug-hauling pilots have been known to operate backwards. That is, they tend to trigger outside the normal range only when the respondent gives a truthful answer. As might be realized from the foregoing, prevaricators are not always drug haulers. The farfetched examples were used for the sole purpose of nailing down one salient point: **ALL DRUG SMUGGLING PILOTS ARE LIARS.**

A private pilot reported he was forced to ditch a rented, single-engine airplane two to three miles off Florida's east coast when the engine malfunctioned while on a flight from Boca Raton, Florida to Atlanta, Georgia. After the ditching, he claimed he used an empty cooler chest for flotation while swimming ashore. According to the pilot, he went to the residence of a friend who lived nearby and waited a day before reporting the accident.

From the very outset, Safety Board investigators assumed a skeptical attitude. The pilot hadn't been too original in his attempt to dream up some reason for the airplane being missing. Air Safety Investigators hear similar accounts when drug-smuggling pilots have their airplanes seized by officials in foreign countries, or when they are involved in accidents outside the United States. Several days after the pilot gave his account of the ditching, aviation authorities in Jamaica reported the airplane had crashed on the island. The accident occurred a day before the reported ditching while the pilot was attempting to take off with a load of marijuana. The pilot abandoned the scene and somehow managed to get off the island without being apprehended.

- -

A pilot reported he had to make a night forced landing in a remote, uninhabited area of South Florida. He said he thought he selected a road but, upon landing with the gear down, the airplane was damaged when it swerved off the road after colliding with some crates and low-growing brush.

When I arrived at the scene, sheriff's deputies, customs agents, drug enforcement personnel, and the pilot were present. The crates the light, twin-engine, airplane's wing collided with were bee hives. Shattered automotive safety glass was strewn along the path, and a fairly new rear-view mirror from a pickup truck or van was found. Paint smears on the mirror's frame matched the color scheme of the airplane. Marijuana residue was found in the cabin and the odor of marijuana was discernible.

It was obvious the pilot was on a drug smuggling flight. The truck was probably used to transport the illicit cargo away from the scene before the pilot reported the accident. It was also believed the truck was positioned to mark, or otherwise identify, the landing site. The operation went awry when the airplane collided with the truck during the landing.

Despite all the evidence, the pilot maintained a cocksure attitude while persisting with his story. Apparently, he was of the opinion he would be subjected to no adverse actions because they had managed to whisk the marijuana away. My role concerned investigation of the accident from an aviation standpoint. The disposition of the illicit nature of the flight was the responsibility of the other officials present.

Some owners and operators attempt to save face by playing it straight when their airplanes are involved in accidents while hauling drugs. Most citizens are aware law enforcement agencies are authorized to confiscate property used to transport contraband cargo. When law enforcement personnel arrive before the drugs are off-loaded, there is not much defense. Our courts consider that prima-facie evidence. In such instances, owners or operators invariably claim their airplanes were stolen. In cases where the cargo is removed before authorities arrive, they dream up legitimate purposes for the flights.

Years ago, we had a rather well-known figure who possessed the certificates required to operate his fleet of Lockheed Lodestars and other airplanes in commercial operations out of his south Florida base. In view of the number of drug hauling occurrences we investigated involving his aircraft, it was considered the certificates authorizing commercial operations were being used as a front for illicit drug smuggling flights. I investigated two drug-related accidents involving his airplanes. Like myself, he was a former Marine Corps aviator, and we had interesting chats during the course of my investigations.

The first accident I covered occurred in a real estate development tract that had gone belly up in South Florida. The airplane was landed on one of the 20-foot-wide paved streets. The landing went well until the left wheel ran off the pavement. Thereafter, the airplane veered left and nosed up in the soft sand. The accident was unreported until found by local authorities several days later. Customs personnel reported they found marijuana residue in the cabin. The operator disregarded the evidence suggesting a smuggling operation. Instead, he said they executed a forced landing following malfunction of one of the engines while on a training flight from an island in the Bahamas to Tampa, Florida.

The second occurrence I investigated involved one of their airplanes that crashed while attempting to make an early morning landing at a fog-shrouded airport in central Florida. Authorities responded immediately and found 2,400 pounds of marijuana in the cabin. Both pilots abandoned the wreckage. The copilot was apprehended but the pilot escaped.

I remember giving the operator my most quizzical expression while he related how the airplane had been stolen off their ramp. Before departing, I handed him an accident report form they were required to submit, while advising him we already knew what had happened. That was the last time I saw him. A few years later he was killed while piloting his WW II fighter plane.

THE ACCIDENT THAT NEVER WAS?

The news story below appeared in the December 3, 1982 edition of The Florida Times-Union, Jacksonville's daily newspaper. The article was written by Joe Caldwell, a Times-Union staff writer, who was one of the most accurate reporters of aviation accident matters I ever met. The article was prompted by the Safety Board's probable cause determination made during a public session because of the wide publicity the accident had received, especially in the Jacksonville, Florida area. As the Investigator-in-Charge, I was present during the proceedings and gave extensive testimony to the Board Members' queries about the factual evidence obtained during the course of the investigation.

Board's Probe Was A Farce, Navy Pilot Says

Lt. Cmdr. V. Allen Spicer, the Jacksonville-based Navy pilot who the National Transportation Safety Board says used bad judgment in ditching a rented plane 17 months ago, is calling the Board's investigation "a farce" and is questioning its field investigators' competence.

Spicer, reached by telephone yesterday in Sigonella, Sicily, where his Navy patrol squadron is on extended duty, said he has obtained a physician's statement backing up his account of the extended exposure to sun and sea after the nighttime landing in the ocean off Ponte Vedra Beach.

"I think that the National Transportation Safety Board's investigation was a farce," he said, referring to the month's-long probe of the circumstances surrounding the ditching of the Piper Arrow airplane on July 8, 1981.

The Board ruled this week that the pilot used poor judgment when he intentionally ditched the plane.

Board Chairman Jim Burnett said, after the Board had ruled, that physical evidence did not support Spicer's statement that there was fire in the engine cowling and that Spicer's explanation of what happened before and after the ditching posed a "credibility problem."

The finding was based on work done by Tom Watson of the Safety Board's Miami office and the Board's authorized "human factors" investigator, Dr. H. Curtis Benson, a Jacksonville Beach physician who is also a pilot and member of the Naval Reserve.

Watson told the Safety Board during its meeting this week that his investigation uncovered no evidence of fire anywhere on the plane when it was pulled from the water 36 hours after the crash. (The airplane was actually found 36 hours after the ditching and salvaged three days later.)

Spicer, who was found on the beach 10 miles south of the ditching site and 56 hours after the incident, said he had decided to ditch the plane when he saw smoke and a flash of fire come from the plane's engine cowling.

"I think Tom Watson has too much of a workload or that he's incompetent because he deliberately misquoted me in the official report," Spicer said.

He said Watson failed to conduct the investigation in accordance with the Safety Board's guidelines and with federal regulations.

"I think that Curtis Benson volunteered for this investigation because he's a publicity hound, and I think he was unqualified as a human factors type medical person," Spicer said.

"I think he far exceeded the scope of his authority and that he also failed to properly analyze my medical information. The statement of Benson that I had no beard growth, no sunburn and lesions in the mouth or ears is incorrect," he said.

"I have gone to a nationally known college of medicine and shown my medical records to a noted doctor who says that what I say happened is backed up by my medical records. I have a letter from the doctor to that effect. He said I was lucky to be alive."

Spicer declined to reveal the name of the school or the doctor, but said he would later.

Benson, reached late yesterday, said the information he passed along to Watson came from physicians who cared for Spicer.

"I merely passed on the information I received from them by direct communication by way of interview." the doctor said. "I was requested by Mr. Tom Watson to aid in his investigation because I was a designated senior Federal Aviation Administration medical examiner and air crash investigator for the Federal Aviation Administration."

Benson said that "at no time" did he initiate contact with the media, adding; "I allowed the media to poke around with me to see what I was doing."

"It was Spicer who got the ball rolling by calling a press conference. It was my feeling that, since this case had already attracted a fair amount of notoriety and publicity and that with our 'Government in the sunshine' [law] it was only fair to allow the media – at their request – to see what we were doing in the investigation."

The physician said that during his interview with the pilot, Spicer made several statements that he said he found difficult to believe.

"These included the statement that he was 350 yards off the beach and was unable to see any lights on shore," said Benson, who re-enacted part of the ditching by going into the ocean close to the spot where Spicer's plane went down. Benson said lights on shore were clearly visible.

Benson, a flight surgeon who has had training in air crash investigation and in environmental medicine, including weather exposure of downed pilots, said: "The scenario as recited by Lt. Cmdr. Spicer taxes my sense of reason."

Commenting on the Board's meeting this week in Washington, D.C., Spicer said: "The whole thing has been like a trial without a jury in that I was never allowed to participate in the investigation, nor was my designated representative allowed to say what he wanted to say."

"I requested [that] the hearing be delayed because I was out of the country, and I wanted to attend to see how these people operate."

Spicer, who flies anti-submarine surveillance P-3 Orion planes for Jacksonville Naval Air Station-based Patrol Squadron 49, said he had filed a complaint, "regarding the conduct of the field investigation" by the Safety Board with U. S. Senator Sam Nunn, D-GA., and U. S. Rep. Charles F. Hatcher, D-GA.

Based on the pilot's statements to the reporter, it's fairly obvious he didn't think much of my competence as an investigator, the investigation that was conducted, Dr. Benson's medical qualifications, or the Membership of the Safety Board who made the probable cause determination. No one is ever "on trial" during the Safety Board's deliberations and there are procedures whereby any person, group, or organization can submit additional evidence the Board did not possess, as a petition for a reconsideration of the Safety Board's findings and conclusions. Safety Board regulations state: "Accident investigations are never officially closed but are kept open for the submission of new and pertinent evidence by any interested person."

These were the pilot's options. One might ask why he didn't submit to the Safety Board a copy of the letter from the noted doctor at the nationally known college of medicine who substantiated, "that what I say happened is backed up by my medical records?" His contemptuous attitude towards the investigative proceedings and the Safety Board's probable cause deliberations bring to mind the testimony of the elderly Faulkner sisters in the movie classic, "Mr. Deeds Goes To Town". During the sanity hearing, the plaintiff's attorney asked: "Tell the court what everybody at home (Mandrake Falls) thinks of Longfellow Deeds." They responded: "They think he's pixilated." After Mr. Deeds took the stand in his behalf, he asked Jane Faulkner: "Who else in Mandrake Falls is pixilated." And Jane replied; "Why everybody in Mandrake Falls is pixilated, except us."

While on standby duty at my residence in Miami, Florida I was notified by the FAA of the accident. They reported the pilot of a single-engine Piper airplane radioed he was ditching in ocean waters off St. Augustine, Florida, shortly after midnight. The U. S. Coast Guard had been notified and their search and rescue procedures were activated immediately.

Our procedures didn't require any immediate actions, and I delayed until arriving at our office the next morning before establishing liaison with the FAA's flight standards district office and flight service station in Jacksonville to obtain the information they had regarding the flight. Thereafter, in keeping with our normal procedures for such occurrences, no further actions were taken pending the outcome of the search efforts for the airplane and the pilot.

My supervisor called me into his office in the middle of the afternoon and told me to travel to Jacksonville the next morning. The occurrence was drawing much attention and our phones were tied up by news media personnel and others seeking information. The accident turned out to be what investigators refer to as a real "head scratcher." Under normal circumstances, an investigator spends two to four days at the scene to cover a general aviation or small business airplane accident. For this nonfatal occurrence in which the small general aviation airplane sustained very little damage except for immersion in salt water, I spent ten days at the scene. My supervisor was briefed daily on the status of the investigation with what turned out to be a pretty trite phrase before I finally returned home: "We still have all the questions and very few answers." His responses were very similar: "Things are fairly quiet in the office, just stay up there and keep churning the pot to see if you can't get to the bottom of the matter."

The airplane was partially suspended in an inverted attitude in ocean waters about 1,000 feet off the beach. It was resting on the tops of the engine cowling and the fuselage above the cockpit. After righting it, the salvage operator floated it to the surface with airbags. Arrangements had been made to bring it ashore at an exclusive beach club that provided ready access to roadways. The landing gear was extended, a long tow rope was attached, and, with the help of many Sunday afternoon bathers, the airplane was towed onto the sandy beach.

The only obvious damage to the airplane were the scratches and abrasions on top of the engine cowling and cockpit on which it had been resting. The initial examinations of the aircraft, engine, and other components were conducted on the spot because of the known adverse effects the atmosphere has on aircraft components that have been submerged in salt water. Additionally, we were aware that some measure of disassembly would have to be accomplished before it could be transported to a salvage yard over public roadways.

I was the only witness at the public session during which Safety Board members reviewed, discussed, sought clarification, and considered the evidence in my Investigator's Factual Report. Thereafter, they voted unanimously for the probable cause determination.

This wasn't a difficult occurrence to analyze, although some loose ends were necessarily left untied because of a lack of sufficient evidence. The pilot's account of the occurrence and other pertinent evidence from my factual report are reported below. Additional information has been included where the reader might require further clarification in order to fully comprehend some data.

The pilot ditched the airplane in ocean waters about 1/4 mile off the beach east of Jacksonville, Florida, at 12:04 a.m. on a Wednesday morning. There were no witnesses to the occurrence and the exact location of the ditching was unknown until the wreckage was found submerged in shallow waters on Thursday afternoon. The fate of the pilot was unknown until mid-morning on Friday when he was found on a lonely stretch of beach about ten

miles south of the ditching site. The pilot was uninjured in the ditching, but was hospitalized several days for dehydration and exposure. The wreckage was salvaged on Sunday, the fifth day after the ditching

The flight had departed the Craig Municipal Airport in Jacksonville on Tuesday at about 10:00 p.m. on a local dual instruction flight to provide an initial checkout in the airplane for the active duty U. S. Naval aviator. After about an hour and a half of dual instruction, the flight returned to the airport to deplane the instructor pilot. At 11:39 p.m., the pilot who had received the checkout commenced taxiing for a local flight to "get the feel of the airplane."

The instructor pilot stated that the pilot told him he wanted a night checkout because he planned to rent the airplane the following evening to take his wife to dinner at Cedar Key, Florida. When the instructor pilot deplaned after the dual portion of the flight, the pilot told him he'd be flying down towards St. Augustine and back on the solo portion of the flight.

The substance of the pilot's account of the solo flight, the ditching, and his survival, included the following: After takeoff, I headed down the coast. Going down, I was not quite to St. Augustine when I heard some type of knocking. It seemed like it was coming from the starboard side of the engine. I decided to just turn around and head back to Craig (airport). I didn't know what it was. Everything in the cockpit, that is instrument indications, looked normal. I was at 1,500 feet, flying VFR (visual flight rules) maybe a mile offshore. I think I had about 2,400 r.p.m., had leaned the mixture, and was using

21 to 22 inches of manifold pressure. Then I thought I smelled fumes, this was a few minutes later. I checked the fuel pressure gauge and it was fluctuating. It dipped into the yellow, still fluctuating, and I turned the electric boost pump on. It was still fluctuating, and I went mixture rich and reached down and changed the fuel selector to another tank. When I turned, I saw a flash out of the corner of my eye and I thought there was flame in the cockpit.

Going back just a second, when I got the fluctuating pressure and fumes, I called flight service just to say, hey, I've got this and am getting back to Craig. When I got the flash I got a rough runner and I just, you know, I thought I had a fire. So the first thing that came to my mind was, you know, get this thing down. St. Augustine being behind me was, I thought, closer but there were no lights at the St. Augustine Airport, they were out for some reason. So again, with this rough runner and this fluctuating pressure and I'm scared to death of fire, I decided, hey, I've got to get this thing down.

Initially, I thought about putting it on the beach and I think I called back and asked for the altimeter setting. I remember the guy coming back and saying, "Can you make the beach?" I didn't know if it was high or low tide, I wasn't familiar with the shoreline and I didn't want to hurt anybody else. I didn't want to lose control of the aircraft and go rolling into somebody's house or whatever. I figured if there was fire the best thing to do was put it in the water and put the fire out. I sideslipped the aircraft from 1,500 feet to get in close to the shore. I wanted to be far enough offshore to get away from the breaking waves, I'm guessing I was about 1,000 feet offshore.

As I was side slipping the aircraft, I had a lip of flame out of the cowling and I chopped the mixture. (Readers should understand that, for all intents and purposes, "chopping the mixture" shuts off all power from the engine.) I kept 90 knots in the aircraft until I got to 300 feet on the barometric altimeter. I was probably doing between 1,000 and 1,500 feet per minute descent, I broke my descent and set 100 feet per minute on the vertical speed indicator. When I chopped the mixture the prop was still turning out there and I didn't see anymore flame. At 300 feet I set a 100 (feet per minute descent) and slowed to 75 knots.

In this aircraft you have the automatic type landing gear that will extend at a certain speed and I remembered this and kept my airspeed until I took care of this because I didn't want to ditch with the gear down. I'm ditching to the north, parallel to the coast, and the moon is setting maybe 30 degrees to port. That's what I'm using more or less as a visual reference, we had about a quarter moon. Like I said, I set 100 feet on the vertical speed indicator, that comes from the U. S. Navy, and the aircraft touched down tail first because I wanted to keep the nose up as long as I could because I think the prop was still turning. The aircraft touched down and skipped, touched down again and skipped, then settled in the water.

Training in the Navy, you know, sit there a couple seconds, let your mind clear. I turned the master switch off and water is beginning to come in the aircraft very fast. I reached over, still strapped in, there's a latch topside and a latch forward on the door, and I couldn't budge it. I got a survival knife out of my briefcase. The aircraft is settling nose heavy, I don't know the degrees down it was but I knew I was down and the water is coming up in the aircraft. I released my seatbelt and am over trying to get the door open and I can't get it, banging it with the knife. I'm not sure but I think the aircraft is hanging nosedown in the water and I'm under water. I figured, hopefully, there's got to be some air up and I was running out of breath. I crawled across the seats into the back of the aircraft and sure enough, I found some air. I got some air and then made two or three trips back for air before I finally got the door open.

My briefcase was on the seat on the passenger side and I shoved it outside with the intention that if it would float, maybe I could hang on. I got to the surface and somewhere in the evolution of getting out I dropped the knife and I couldn't find the briefcase. I decided I'd better get my shoes off and as I reached down to get my shoes I remembered, maybe these things will float, and I remember emptying the water out of my shoes and sitting them on the surface. I may have been out of the airplane a minute or five minutes, I was totally exhausted. I was doing a drownproofing method to rest and that's when whatever it was came by and hit me on the leg and moved me in the water. The first thing I thought of was a shark and I remained still as possible while doing a slow motion drownproofing. I knew my clothes were extra weight and I'm tired. I took my pants and shirt off and tied my pants leg and shirt together. I did that because I had on a white shirt and light colored trousers and I knew I wasn't much of a target out there. I held onto my pants with my left hand. I had taken the billfold out and stuck it down in my jockey shorts.

I continued drownproofing and don't know how much time elapsed but I saw a helicopter. First I heard it and then I saw it. My vision at night could be off but I'm guessing the helicopter was 300 to 500 yards away, fairly low. I'm guessing they'd be doing a ladder type search, back and forth, but evidently they didn't see me and I didn't see or hear anything else that night. I released my trousers. When I got out of the airplane I was tired and I made my mind up right then that I had to conserve energy. I've got to stay alive. I know they are going to be looking for me and I know they're going to find me. I just told myself that.

I committed myself to a relaxed drownproofing method so that I could rest as much as possible. The next day comes and I don't know how far from shore I am but I can't see land. I want to think on the afternoon of the first day I saw a P3 aircraft, (Navy four-engine patrol plane) but again I knew I couldn't be wasting the energy to be waving. It was a fairly calm sea until late that afternoon when a hellacous thunderstorm came up. I know the thunderstorm gave me extra strength because I was able to get possibly two glasses of water by cupping my hands under my lips. Nightfall comes and I know they're still looking for me. Of course, there's no way of measuring time, I kept my eyes closed as much as I could because I knew salt water would cause damage. I didn't see or hear anything that night.

The next day was Thursday and I'm still doing my drownproofing, that's what I did the whole time. I never tried to swim, I wanted to save that energy. I opened my eyes and saw three sharks. I didn't know if they were interested in me. I yelled in the water several times and they finally ambled off. Its getting late afternoon, I'm very tired and know I can't last much longer. I knew I had to do something and I shoved myself out of the water and I saw land. My vision was very blurred and I couldn't tell how far away it was but I knew it was land. I tried to orient my motion towards the land and I'm still doing my modified drownproofing with a little bit of a kick in it.

I have no idea when the tides are but I have in mind that if I can see land now and I couldn't before, the tide must be bringing me in. Nightfall comes and I'm not to land yet. I pick out a light maybe 30 degrees starboard from where I was in the water so I knew I hadn't been imagining this and I'm thinking, I'm going to make it. I get in close enough to hear the waves breaking. I know I've got water in my ears and everything and its probably cut down on my capability of hearing but I figured I was fairly close so I give it all I can and my foot touches bottom. Again, I don't know what time it was. I have no concept of time but I figure it was sometime in the middle of the night, maybe between 11:00 and 2:00 in the morning and I just, you know, collapsed there on the beach.

The pilot was found lying on the beach by a passerby shortly after 10 a.m. on Friday, which was about 58 hours after the ditching.

The statement by a specialist on duty at the Jacksonville Flight Service Station included the following: (Reported times were found to be several minutes fast.) "At 12:08 a.m., N29312 called on Frequency 119.7 and advised me that runway lights at St. Augustine did not appear to be on. After this transmission the pilot immediately stated that he was east of the St. Augustine Airport, had fumes in the cockpit and the fuel pressure was fluctuating. At 12:09, the pilot advised that he had a fire in the cockpit, requested the Craig altimeter setting, then stated, 'It's getting hot in here, I'm going to put it down,' and stated he was off St. Augustine Beach north bound. I issued altimeter setting. No further contact with the aircraft.

I immediately notified Jacksonville TRACON, then called the Coast Guard at Mayport, Jacksonville's Air Route Traffic Control Center and General Aviation District Office, and then the St. John's (county) Sheriff's Department."

The airplane was equipped with an altitude encoded transponder that signals its position and altitude. The Jacksonville Flight Service Station was not equipped to receive the transponder signal; however, investigation showed the transponder return signal was received and automatically stored in a computer at the FAA's Jacksonville, Florida, Air Route Traffic Control Center. Readout of computer data showed an aircraft southbound off the beach east of Jacksonville at 1,400 feet altitude at 11:51 p.m. The aircraft continued southward about ten more miles before executing a full 360-degree circle. Thereafter, it again proceeded southward down the beach about four more miles before reversing its course and commencing a descent. The final position of the aircraft recorded by the computer was at a point about three miles south of the wreckage location while the flight was at 400 feet altitude. The actual ditching site was about 1/4 mile off the beach in the vicinity of the point where the aircraft had executed the full 360 degree circle.

The weather at the Craig Airport was clear with seven miles visibility and calm winds at the

time of the ditching. Satellite photographs obtained from the National Hurricane Center in Miami, Florida, showed fairly clear skies in the vicinity of Jacksonville until noon on Wednesday when some cloudiness began to form that persisted for the remainder of the day. The photographs for Thursday showed skies in the Jacksonville area to be relatively free of cloudiness and those that did form, showed a thin appearance.

After the pilot of a light civilian airplane spotted the wreckage about 11 miles southeast of the Craig Airport, divers from the U. S. Naval Station, Mayport, Florida, proceeded to the scene. Upon diving on the wreckage, they reported the pilot was not on board and the aircraft appeared to be intact. The cabin door was missing and they were unable to find the door in the vicinity of the wreckage.

It was noted that there was little or no movement of the wreckage between Thursday, when it was found, and Sunday, when it was salvaged. The airplane sustained no appreciable damage during the ditching. Examination showed no evidence of in-flight fire in the engine compartment, the cockpit/cabin area, or elsewhere on the airframe. Salt water erosion of the magnesium components on the engine precluded a test run. On-site examination of the engine, and subsequent disassembly, showed no evidence of in-flight failure or malfunction.

The hinges on the outside of the cabin door were of the clevis type and the door was attached by a pin and cotter key. The pins and cotter keys were missing but the door hinges showed no visible distortions. There were scratch marks on the fuselage above and below the upper door hinge. The scratch marks were 9/16 of an inch apart. It is

noted that the opening on the clevis hinges were only 6/16 of an inch apart. The operator reported the scratch marks were not on the doorframe when the flight was initiated. Under the scenario reported by the pilot, the marks wouldn't have been made during the ditching. The scratches were similar to those that might have occurred if a prying action had been applied after inserting a pronged tool around the hinge.

The pilot was admitted to the Flagler Hospital in St. Augustine through their emergency room after being found on the beach. Dr. Benson's statement following his review of the pilot's medical records included the following:

"Certain aspects of the data reviewed are consistent with some dehydration and starvation. However, the lack of sunburn on the exposed parts of the pilot's body is inconsistent with the exposure which he describes. Also, the lack of beard growth causes some concern. Additionally the pilot's vital signs are not consistent with significant dehydration upon admission to the Flagler emergency room. Specifically, a pulse rate of 92/min. with a blood pressure of 132/90 does not suggest a significant volume depletion. Furthermore his Hemoglobin and Hematocrit do not show evidence of significant hemoconcentration. They are essentially the same taken several days later as they were upon admission to Flagler Hospital. The differential count of the white blood cell series is likewise normal which is inconsistent with a stress reaction. E.N.T. examination failed to show any salt water lesions on the lips or in the mouth."

In preparation for my testimony before the Safety Board, Dr. Benson was asked to explain in

lay terminology the significance of some pertinent points. He replied that our blood is made up of about one-half blood and one-half other liquids, and when you have significant dehydration, the other liquids are depleted, resulting in hemoconcentration, which did not occur in this instance. He further explained that a person's normal reaction to stress causes a significant rise in the differential count of the white blood cell series, which did not occur in this instance. He also noted that a person who spends 48 hours in salt water normally shows salt water lesions or sores on the lips and in the mouth.

A request was made to the airplane manufacturer's representative for a flight test in a similar model airplane to determine the rate of descent with the power off and the landing gear and wing flaps retracted. Subsequent to the flight test, they provided a graph showing a rate of descent of 1,155 feet per minute at 90 knots indicated airspeed and 880 feet per minute at 75 knots.

Remember, the pilot said after he reached 300 feet altitude, he slowed to 75 knots and maintained a 100-feet-per-minute descent rate. On three separate occasions in his statement, the pilot made reference to the 100 feet-per-minute rate of descent. Well, that's impossible, because he had already "chopped the mixture" which essentially shut off all engine power. Thereafter, the pilot was basically in a "dead stick" configuration whereby the forces of gravity and gross weight of the airplane, in conjunction with the indicated airspeed, were the factors that would determine the descent rate. While the pilot can always increase the descent rate by upping the airspeed, the reverse is only true until

the airspeed decreases to the stall speed. Thereafter, further back pressure on the control stick or control column in an attempt to maintain a 100 feet per-minute-rate of descent can result in disastrous consequences when the airplane stalls.

After arriving in Jacksonville the day after the ditching, I visited the Naval Air Station and extended an invitation to officers in the pilot's squadron for their participation in the Safety Board's investigation. The officers I conferred with declined the invitation.

The above data includes the pertinent information contained in my official report which the Safety Board refers to as the "Investigator's Factual Report." It also encompasses the significant data the Safety Board's presidentially appointed members considered in their deliberations. There were some other interesting aspects to this investigation; however, before proceeding further, why not ponder in your own mind the data that has been presented and determine whether you are willing to accept the pilot's statement in toto, or whether you agree with the Chairman, other Board Members, and the investigative team, that it poses a "credibility problem."

The Naval officer who headed up the Explosive Ordnance Disposal (EOD) unit that came out from the Mayport, Florida, Naval Station to dive on the wreckage was one of the most motivated fellows I ever met. He tackled his assignments with gusto and enthusiasm. With his handsome features and muscular build, he looked as if he belonged in the movies. It was obvious the members of his crew had a great deal of respect for him. We were engaged in chitchat after he completed his assignments and part of the dialogue went something like this: He said, "If you'd asked me, I'd have told you what I told my crew after we finished searching the ocean floor for that cabin door." Since it was my practice to listen to whatever anyone thought or had to say about my investigations, I responded, "That's what I want to know, what'd you tell them?" He said he told them he thought it would be wise to check the pilot's insurance policies, because there isn't a single soul who wouldn't have expended every effort attempting to swim such a short distance to the shore. He stated he had fully expected to find the pilot in the airplane, because not attempting to save himself by swimming ashore just went against human nature and our basic self-preservation instincts.

The photographer who took motion picture footage of the pilot while still lying on the beach, and again, as he was wheeled out of the emergency room at the Flagler Hospital in St. Augustine, was interviewed. I asked about his impressions of the pilot and whether he gave the appearance of someone who had been exposed to the elements for 58 hours without food and water. He answered with an emphatic "NO," stating the pilot looked to him like someone who first made crawl marks on the beach, then covered himself with sand and waited to be found.

Mr. Mel Martin, the news director at the NBC television station in Jacksonville, was observed to be another of the media folks who never attempted to distort the facts. His station possessed the motion picture footage, and the assistance he rendered by permitting Dr. Benson and myself to review the newsreel coverage over and over again was invaluable. We were able to stop the reel where appropriate and take still shots of the television screen. We utilized this procedure to obtain excellent close-up photographs of the pilot lying on the beach, and as he was being wheeled out of the emergency room at the Flagler Hospital.

The circumstances of the pilot's exposure to the elements were discussed with the FAA's regional flight surgeon in Atlanta, Georgia. The flight surgeon remarked that he would have expected the pilot's shoulders to be blistered from such an ordeal.

The factual information and medical data were discussed with Dr. Joseph H. Davis, the Dade County, (Miami) Florida, Medical Examiner and a Professor of Pathology at the University of Miami. Dr. Davis called it "a most fascinating investigation." An excerpt from the doctor's response was as follows:

"The photographs do not suggest a 58 plus hours' growth of beard. He has dark hair and a mustache, so a beard in excess of 58 hours ought to be more visible than revealed in the photographs. Furthermore, his facies do not suggest significant dehydration. There appears no skin lesions to suggest salt and sun exposure."

After his review of the factual evidence, Dr. Davis posed the following questions:

"Why claim fire when there is none? Why swim around 48 hours a quarter mile from shore? How does a door break off when there is no break?"

The pilot was reluctant to release his medical records when they were requested. Although I was aware that the Safety Board had authority to subpoena the records, the pilot was informed we would wait a few days for his decision. I then employed a practice sometimes used in such situations. Investigators receive numerous calls from news media personnel throughout the conduct of such investigations. When they asked about the status of the investigation, I replied, "Well, right now we're waiting for the pilot to decide whether we can have access to his medical records." I knew they would put it on the air or publish it in the newspapers, and figured the Navy brass might question why the pilot was reluctant to release his records. I don't know whether my tactics had any effect, but during my next conversation with the pilot, he authorized us access to his medical data.

In some instances Dr. Benson, a Navy Reserve Captain, was said to be a hired human factors consultant to the investigation. Dr. Benson wasn't hired nor did he receive any remuneration whatsoever for his services. I first met him while documenting the wreckage after it had been moved to an auto salvage yard. Upon learning of his background and qualifications, in conjunction with the keen interest he exhibited towards the proceedings, I asked whether he would be willing to participate in the investigation. Specifically, he was asked to obtain and review the medical data. Dr. Benson readily accepted, and I have rarely seen anyone who tackled his tasks with such zestful enthusiasm. Somehow, he was able to rearrange his office schedule in a manner that permitted him to give the better part of three or four days to the Safety Board's investigation.

He also obtained use of a boat, jumped overboard at night near the ditching site, and found that lights on shore were clearly visible. He performed the experiment for the sole purpose of determining whether he could see lights on shore, because the pilot had told us he couldn't see any lights after escaping from the wreckage. Dr. Benson also obtained use of an airplane and made flights with the cabin door on, and with the cabin door off, to determine whether flight service station personnel could notice any significant change in his voice transmissions. The results of this experiment were inconclusive, but I looked on it with the attitude, "nothing ventured, nothing gained." Additionally, he performed the human factors aspects of the investigation in a most professional and competent manner, and the Safety Board is extremely grateful for his services.

Safety Board investigations are not legal proceedings, as such, and the facts obtained do not have to stand the rules of evidence tests that might apply for matters under litigation. Safety Board investigators are encouraged to obtain evidence from any source considered reliable. Likewise, the Investigator-in-Charge is authorized to name as parties to an investigation any persons or organizations having information or expertise that will be required. Early in the proceedings, the pilot had questioned the overall scope of the Safety Board's investigative authority. In response to his request, he was permitted to borrow my investigator's manual, which was returned by the pilot's lawyer several weeks later. Contrary to the pilot's claim, the investigation was accomplished in accordance with the Safety Board's basic guidelines and modus operandi.

Basically, the Safety Board is concerned with the aviation aspects of an occurrence regarding the pilot, airplane, controller, briefer, mechanic, etc. The investigation into any criminal, unlawful, suicidal, or ulterior motives that may be involved in an occurrence is normally relegated to other responsible agencies, organizations, or bodies. This precept was involved in the sole finding by the Safety Board that the pilot exercised poor judgment when he elected to ditch the airplane in this instance.

A first anniversary update on the ditching appeared in the Florida Times-Union. The article, by staff writer Joe Caldwell, elicited a vehement response from the pilot's commanding officer. His letter to the editor was printed on the editorial page of the July 16, 1982 edition. That was five months before the Safety Board released its findings. In the interest of presenting opposing views, if that's what this might be called, the commanding officer's letter follows:

Spicer's Commander Says Story Outrageous

by Cmdr. J. L. Minderlein

As Lt. Cmdr. V. Allen Spicer's commanding officer, I feel that it is my responsibility to comment on your article of July 4 in reference to the anniversary of that aircraft accident. As a friend, I am outraged by what I regard to be sensational and irresponsible journalism.

The article offers no new information, is filled with innuendos and is nothing more than speculation at best. To be specific, Times-Union staff writer Joe Caldwell admits that the National Transportation Safety Board has yet to announce its conclusion after conducting a full-scale inquiry. At the same time, he makes a definitive statement that "no evidence of either fire or a mechanical malfunction could be found."

Caldwell alludes to the fact that the incident may have been an intentional act for whatever reason. If this was not a bona fide accident, then I challenge Caldwell to provide "whatever" reason for a man to make a night ditching in an unfamiliar aircraft. Only those of us who fly aircraft can appreciate the magnitude of the situation. The average pilot would not have survived this accident. The fact that Lt. Cmdr. Spicer did survive is testimony to his skills as an aviator, his strength of character, a strong will to live, and his Navy survival training. For "whatever" reason would a man intentionally subject his wife and family to the ordeal that I witnessed last year?

Again, Caldwell insinuates some misconduct on Lt. Cmdr. Spicer's part when he says that "the Navy launched its own investigation, since the incident posed some questions about the conduct of one of its officers." Let me inform the public that such an investigation is a routine administrative matter required by Navy Regulations. The Navy launches a similar investigation when a man falls off a motorcycle.

As to the innuendo that Lt. Cmdr. Spicer turned up in "surprisingly good physical condition," I think Caldwell might have mentioned something about Lt. Cmdr. Spicer's vital signs and muscle damage that was incurred by saltwater immersion in addition to the lack of sunburn and beard growth. (I do not know on what basis Dr. Benson made his assessment of lack of beard growth since he did not personally see Lt. Cmdr. Spicer after the accident.) Having been at his side only hours after being found, I can personally attest to his physical condition. I could feel the exhaustion in his voice. If that was an act, then Al Spicer deserves Henry Fonda's Oscar.

It is also not hard for me to understand why Dr. Benson still has questions in his mind about the case. Anyone who thinks he could possibly recreate the degree of panic, confusion, and sheer exhaustion experienced by Lt. Cmdr. Spicer that night by placing himself in the water under carefully controlled conditions for a short period of time deserves to continue to have questions in his mind.

I could go on and on; however, I might sum this up by stating unequivocally that I have no questions in my mind about Lt. Cmdr. Spicer. He is not only an outstanding Naval officer and profes-

sional aviator, but an outstanding individual and family man as well. The trust and confidence that we place in this man can easily be seen since, for the better part of the past year, he has been responsible for the overall training efforts of this squadron and, in particular, for training and certification of our pilots. I respect him and trust his judgment implicitly.

Isn't it a shame that a man who manages to survive a life-or-death experience that most of us cannot even begin to appreciate gets no better treatment from the local news media than distrust and doubt?

In my opinion, the pilot's commanding officer had a severe case of "Reeboks in Choppers Syndrome" when he drafted his editorial reply. Strictly in the interest of providing clarification for some of the issues raised, the following comments seem appropriate:

1. The Safety Board encourages its investigators to release factual information obtained during the course of their investigations. I had personally informed staff writer Joe Caldwell that no evidence of either fire or mechanical malfunction could be found and it was also reported in a Jacksonville newspaper edition during the conduct of the on-scene investigation. The collection of such factual information is the responsibility of the Investigator-in-Charge and not a conclusion normally reached by the Safety Board in its deliberations.

2. As stated previously, Dr. Benson put himself in the water at night near the ditching site to determine whether he could see lights on shore. His purpose was not to "recreate the degree of panic, confusion and sheer exhaustion the pilot experienced," as claimed in the editorial response. Some might question the judgment of the

pilot's commanding officer for speaking so disparagingly of the efforts of a more senior Naval officer who provided so much cooperation, assistance, and expertise during the course of the Safety Board's investigation.

3. The Navy's investigation that the pilot's commanding officer referred to as "a routine administrative matter required by Naval Regulations," included 25 enclosures. A question might be raised; "Is that what the Navy calls routine?"

4. The Naval investigative report contained the statement: "No analysis of the engine was conducted." Well, maybe the Naval officer who conducted the investigation didn't consider the examinations and subsequent teardown of the engine to be an "analysis," but the Safety Board certainly did.

5. The pilot's commanding officer concluded his reply with accolades for Lt. Cmdr. Spicer's performance as a Naval officer and the outstanding character he exhibited as a responsible citizen and family orientated person. Nothing disparaging was ever mentioned about the pilot in either of those arenas in the news media releases, and I believe I read or heard most of them.

Questions were raised concerning the pilot's veracity, and rightly so in the opinion of the investigative team, and ultimately, the members of the Safety Board. If the pilot's commanding officer considered that slanderous of Lt. Cmdr. Spicer's overall character, then sobeit. In our society, it's generally recognized that going to church on a regular basis and joining in little league activities with your kids presents a favorable manifestation of an individual's character. However, our newspapers are replete with articles reporting the circumstances whereby some of our outstanding citizens, with seemingly impeccable character, are, in a manner of speaking, getting caught with their hand in the cookie jar.

A difficult aspect to fathom was the ostrich-like stance of responsible Naval personnel in the Jacksonville area. Apparently, they listened to everything the pilot had to say, bought his story carte blanche, gave him a pat on the back, arranged a press conference on base because they believed the publicity would be favorable, and then buried their heads in the sand. They refused to participate in the investigation, review the pilot's statement for inconsistencies, attempt to determine what factual evidence was being gathered, and obviously failed to conduct their own in-depth analysis of the medical data. However, it is noted that the Safety Board's files in Washington showed Member Engen, a retired U. S. Navy Admiral, provided a briefing for the Deputy Chief of Naval Operations for Air Warfare prior to the Safety Board's deliberations. Accordingly, no attempt is being made to indicate the official opinion or stance of the U. S. Navy regarding this occurrence.

The pilot was the defendant in a lawsuit filed to recover the cost of the airplane. Safety Board investigators are not permitted to testify in court in liability cases; however, when the case came to trial, I had retired. The plaintiff's attorney called me as a witness. The scope of my testimony was limited to the factual evidence obtained during the course of the investigation. Since I had been the Investigator-in-Charge, Safety Board regulations precluded me from appearing as an expert witness or giving opinion testimony concerning the cause of the accident. These restrictions are encompassed in the Board's regulation stating: "The Board relies heavily upon its investigator's opinions in determining the cause or probable causes of an accident, and the investigator's opinions thus become inextricably entwined in the Board's determination," and, "they shall decline to testify regarding matters beyond the scope of their investigation or to give opinion testimony concerning the cause of the accident."

Upon learning that the plaintiff attorney's presentation would be related only to matters concerning the pilot's ditching of the airplane without reasonable cause or justification, I was very skeptical. He was advised no jury would ever return a favorable verdict if all he intended to do was raise questions about the pilot's judgment. He persisted with the trial anyway, and I gave rather lengthy testimony concerning the methods used to conduct fire-related investigations. Thereafter, I testified concerning the examination of the airplane and the fact that we found no evidence whatsoever of fire.

Although I wasn't permitted to attend the remainder of the trial, my wife was in the courtroom. She said after the plaintiff's attorney presented his case, the defense called a Naval officer as a witness, who stated he was on the beach when the wreckage was recovered. He testified that upon removal of the engine cowling, he observed some discolorations on the inside. Please understand that no attempt is being made to discredit the officer's testimony in any manner whatsoever. His observations were correct; however, the inside of the cowling on an airplane's engine is just like the inside of the hood on an automobile. It's never as clean as the outside, but that doesn't necessarily signify that any discolorations resulted from fire.

Nothing derogatory is intended towards our court system either, because, in my opinion, the plaintiff's attorney put on a pretty sorry show that was doomed to failure from the very outset. At the conclusion, the jury returned a "not guilty" verdict. In defense of the plaintiff's attorney, it is realized that his failure to delve into the inconsistencies in the pilot's statement and the other evidence which did not support his assertions, may have been attributable to the relatively small sum (approximately $50,000) being sought.

To onlookers, it might sometimes appear that those involved in an accident and those conducting the investigation are lined up on opposite sides of the street, holding personal vendettas against each other. Under normal circumstances, that's not the case at all. The somewhat adversative nature of the proceedings in this instance occurred when the investigators had to divert from their roles as collectors of evidence and don their detective caps because of what they detected as inaccuracies in the pilot's account. It's probably a natural reaction for people to take offense to those who question their account of some happening. However, the investigators have no axes to grind. They're just people with a job to do, and those with a high degree of loyalty and devotion to duty can become completely engrossed in their assignments.

The parties to this investigation represented the ultimate that could be assembled from a technical standpoint. They included a highly-qualified FAA maintenance inspector, the senior accident investigators for both the airplane and engine manufacturers, and Dr. Benson. If I could name the parties again, not a single person would be excused or replaced. Most folks will readily agree that it's impossible for anyone to conduct such a lengthy investigation without forming some conclusions and opinions of his own. Accordingly, a discussion of the pertinent issues, as I saw them from my position as the Investigator-in-Charge, follows:

The pilot's entire account of the flight, the fire, the reason for the ditching and his survival at sea for two full days is suspect. None of the personnel who participated in the investigation are willing to accept his story without some serious doubts. The statements by the doctors who reviewed the medical records show that they have some reservations as to whether the data is consistent with that which might be expected for a person exposed to the elements in salt water for some 48 hours. A question arises as to whether any shirtless person can remain in a near face-down position on the ocean surface for two full days, one and one-half of which were in relatively bright sunshine, without sustaining severe sunburn.

The possibility exists that the pilot may have ditched the airplane for some ulterior motive. It is not reasonable to assume that a professional aviator would make a decision to ditch an airplane in such close proximity to an airport after seeing a lip of flame out of the cowling. It is not reasonable to assume an individual would allow himself to remain at sea two full days after ditching so close to shore. It is not reasonable to believe an individual adrift at sea hoping to be found would not expend some energy and effort to wave and signal to the only airplane that passed overhead during his ordeal. According to the pilot, he never intended to swim ashore after escaping from the airplane in the middle of the night. One of the first things he said he did was remove his shoes and place them on the surface with the obvious hope that they might assist rescue personnel who would be looking for him. Such actions just don't coincide with the basic will to survive inherent in all individuals. Rational action would have dictated that he make every effort to save himself by swimming the relatively short distance to shore

In view of the circumstances of the occurrence, the full circle the pilot executed while proceeding southbound near the shore merits some consideration. Remember, after completing the turn, he proceeded about four miles farther down the beach before reversing course and returning to the vicinity of the circling maneuver before ditching the airplane. No firm conclusions can be drawn from these events; however, in my opinion, they present data for much speculation and conjecture.

The wreckage examination showed the pilot did not experience a sustained fire. It also raised serious questions as to whether the cabin door was on the airplane at the time of the ditching. In reality, a more appropriate title for this chapter might well be: **"THE CASE OF THE MISSING DOOR."** It must be remembered the pilot's entire scenario comes unraveled if the cabin door was not on the airplane at the time of the ditching. That could raise all kinds of questions, and provides the basis for all manner of conjecture, as to just how he managed to be found on the beach some two and one-half days later.

The airplane was not subjected to any forces that would have torn the door off after the ditching and the entire investigative team was of the steadfast opinion that the door was not on the airplane at the time of the ditching. Additionally, the marks on the fuselage in the vicinity of the upper door hinge are not consistent with anything that could have occurred while the airplane was in the ocean. Obviously, the whereabouts of the cabin door, and a rational explanation of the events that caused the scratches in the vicinity of the upper door hinge, are still without reasonable explanations.

Another area of concern relates to the three occasions during the interview of the pilot when he stated he maintained a 100 feet-per-minute rate of descent after "chopping the mixture," which, as previously discussed, was impossible. The taped interview of the pilot was conducted at his residence with myself, Bill Bookhammer from the FAA, and the Naval officer who conducted the investigation required by Navy Regulations, present. Bill and I had completed our initial examination of the wreckage and already had suspicions because of the lack of evidence regarding fire and a belief held by all parties to the investigation that the cabin door was not on the airplane at the time of the ditching. Both Bill and I noted the inconsistency regarding the 100 feet-per-minute rate of descent as soon as it came out. As we drove away, I remarked to Bill: "If you're willing to buy everything we heard in there carte blanche, I've got some high and dry property a couple miles east of Miami Beach you might be interested in."

So we come to the end of this investigative effort. We know more than we did at the beginning, but we still can't come up with a rational explanation for everything in the pilot's account of the accident. In his second letter to the church at Corinth, the Apostle Paul wrote: "For we must all appear before the judgment seat of Christ, that each one may receive what is due him for things done while in the body, whether good or bad." With a strong commitment to that precept, it is believed the pilot will have to rehash this entire matter on at least one more occasion. At the appointed time, it is considered he will have to come up with a more plausible rendition than he has to date. Otherwise, those residing in the next world probably are not going to buy his story either, whether or not pixilated.

MEDICAL MEANDERINGS

"The autopsy protocol showed no evidence of human factors involvement." That statement usually appears in the section of accident reports entitled "Medical and Pathological Information." Basically, it concerns the general health and well-being of the flightcrew in conjunction with the findings during post-mortem studies. Human factors relating to causes other than drug or alcohol abuse are involved in a number of accidents. Eight of my investigations involved accidents that occurred after the pilots became incapacitated. Additionally, incapacitation was suspected in several other instances where meaningful pathological studies were precluded by the circumstances of the accident.

The statement, "Toxicological studies were negative (or positive if that were the case) for alcohol, drugs, and carbon monoxide" is used in conjunction with that above. Safety Board statistics will show impairment of the pilot's ability and judgment because of alcohol or drug use is the most prominent factor in the human factors arena. As we all know, the more some folks drink, the braver they get. Since alcohol affects an individual's mental state and judgment as well as his coordination, some pilots, after a few drinks, might attempt to fly in unfavorable weather or take other risks they would normally avoid.

- -

FIRE ON BOARD

Some say experience is the greatest teacher. That might be true, but in my opinion, nothing beats a king-sized boo-boo as a source of retained learning, especially when we're trying to save face. On one occurrence, a pilot, flying in near hurricane conditions, crashed in Florida. The accident site was located only 26 miles from the point of departure. The examination of the wreckage showed in-flight separation of the vertical fin and rudder. There was no evidence of powerplant failure or malfunction, nor was there evidence of in-flight or postcrash fire. An in-depth weather study by our meteorologist in Washington showed that instrument weather conditions with low ceilings and visibilities would have prevailed, with the likelihood of severe turbulence being encountered. Since the accident airplane was about the only one in the sky that day because of the existing weather, it was easy to proceed down the primrose path and write this off as a weather-related occurrence caused by a loss of control due to spatial disorientation (vertigo) by the noninstrument-rated pilot.

For reasons that were never fully explanatory, an inordinate degree of difficulty was experienced in obtaining the medical examiner's report. After repeated telephone calls, we finally received the medical examiner's findings. The autopsy protocol showed nothing significant to the accident. However, the toxicological studies showed a blood/alcohol concentration of 0.24 percent and a carbon monoxide saturation of 52 percent.

The wreckage was not found for three days, and the blood/alcohol concentration could have been altered to some degree by post-mortem changes in the biological specimen. However, it was too elevated to be wholly attributable to post-mortem changes, and indicated the pilot was under alcoholic influence to a degree sufficient to produce emotional instability, loss of critical judgment, and some muscular incoordination.

Under normal circumstances, we would have expected to receive a call from the medical examiner's office during the early stages of the investigation as they attempted to determine the precise factors associated with the elevated carbon monoxide findings. Specifically, they would have wanted to know whether the wreckage examination showed evidence of in-flight fire, postcrash fire, or cracks in the exhaust system that could have permitted noxious gases to seep into the cockpit/cabin area.

The relatively high carbon monoxide concentration caused grave concerns and resulted in a flurry of activity on my part to resolve the matter. A rather intensive study of the subject was accomplished, and information was solicited from toxicologists and medical personnel, including those at the Armed Forces Institute of Pathology in Washington. Personnel at the Institute of Pathology forwarded excerpts from the Journal of Forensic Sciences containing the results of in-depth studies of carbon monoxide's toxic effects that were very enlightening. Correlating the data with the short duration of the fatal flight, it was concluded that the carbon monoxide concentration would not have been incapacitating for the pilot. However, in conjunction with the elevated blood/alcohol reading, it was considered the pilot's aeronautical ability was impaired.

Thereafter, efforts were made to determine the whereabouts of the wreckage that had been released at the conclusion of the on-scene investigation. As in many small airplane configurations, the cabin heater was operated by a heat exchange from the engine exhaust gases. The cabin heat control was in the OFF position, but the heater components separated and were not found in the salvaged wreckage or on two revisits to the remotely located accident scene. Extensive disintegration of the aircraft structure had occurred during the crash sequence, and the precise causes of the abnormal carbon monoxide concentration were not determined.

The investigation showed three possible causes for the accident: In-flight separation of the vertical fin and rudder; pilot incapacitation due to alcoholic indulgence and carbon monoxide contamination; and a loss of control by the noninstrument-rated pilot who was attempting visual flight in instrument weather. After a detailed study, it was concluded the vertical fin and rudder separated because of pilot-induced aerodynamic forces in excess of the design limitations. Alcoholic indulgence and carbon monoxide contamination impaired the pilot's ability to control the airplane. However, in view of the relatively short duration of the flight, it was considered complete incapacitation did not occur. Accordingly, the predominating factor in the accident was believed related to a loss of control by the noninstrument-rated pilot who attempted the visual flight in extremely hazardous weather phenomena.

All this activity had consumed an inordinate amount of time. The investigation was conducted shortly after my employment by the Board. My supervisor chalked it up to normal progression on the experience curve. Besides everything learned about carbon monoxide and its effects, the prudence of retaining custody of the wreckage until assured a further examination would not be required, was demonstrated most effectively.

Carbon monoxide owes its toxic properties to the fact that it combines with hemoglobin, the oxygen-carrying component of the blood, to form carboxyhemoglobin (COHb). Hemoglobin in combination with carbon monoxide is not capable of transporting the required oxygen to vital organ systems. Our blood has an affinity or attraction to an existing carbon monoxide concentration in the atmosphere that is about 200 times greater than it is for the oxygen in that air. Accordingly, in those instances where a fire occurs, it is possible to rapidly accumulate a high carbon monoxide concentration that will linger because the elimination rate is much slower than its uptake. Carbon monoxide causes all manner of undesirable symptoms including headaches of varying degrees, dizziness, nausea, etc. Concentrations above 50 percent can result in coma with intermittent convulsions, while higher levels can cause respiratory failure and death.

Individuals can only introduce carbon monoxide into their blood stream by breathing contaminated air. One of the purposes of carbon monoxide studies is to determine whether victims died in the crash, or as a result of an in-flight or postcrash fire. Concentrations of 10 percent or less are normally disregarded because heavy smokers can attain that level. Victims having concentrations above 10 percent are usually considered to have survived the impact forces.

While flipping through the TV channels one day in 1989, the selector stopped on a program where a Canadian official was being interrogated about the McDonnell-Douglas DC-8 accident at Gander, Newfoundland. The airplane crashed on takeoff, killing 248 American servicemen. The Canadian official was responding to claims by the next of kin that some of the victims who died instantly showed extremely elevated carbon monoxide concentrations. Basically, the Canadian investigator dismissed the claimant's stance with something akin to: "Well, the investigation showed no evidence of in-flight fire and the elevated carbon monoxide findings weren't fully explanatory." No attempt is being made to align myself with either side of the issue. However, if a number of victims killed instantly showed abnormally high carbon monoxide concentrations, it is reasonable to assume there had to be an in-flight fire on the airplane.

The FAA requires that the fixtures and furnishings in the cabins of passenger models meet various fire-resistant or fire-retardant standards. One of my investigations involved a jet transport model that caught fire while the flightcrew were conducting their initial preflight checks before any flight attendants or passengers were boarded. The copilot boarded the airplane after plugging in the external electrical power source. After checking the rear of the cabin, he went forward to the cockpit. Upon looking back he saw smoke coming from over one of the rear seats. He returned to the rear of the plane just as the captain came on board. When the captain saw the smoke, he picked up a fire extinguisher and joined the copilot. Upon discharging the fire extinguisher underneath the overhead rack, flames shot out the top. When the extinguisher was discharged into the area on top of the overhead rack, more flames shot out the bottom. Thereafter, both crewmembers abandoned the airplane, the airport fire department responded immediately, and the fire was extinguished within ten minutes.

The damage to the cockpit and cabin was unbelievable. The soot-coated material hanging from the ceiling and overhead racks resembled stalactites on the roof of a cave. The seat cushions and all interior furnishings were melted, burned, or consumed in the fire, and the entire cabin and cockpit showed a heavy soot coating that completely obscured the cockpit instrument readings. From a safety standpoint, this was considered to be a real jewel. That's because we had been led to believe such massive destruction in so short a time frame was not possible. However, all our efforts to direct attention to the destruction for the purpose of requiring more fire-resistant materials for cabin interiors were to no avail.

The problems with raising even an eyebrow to the inadequacy of the interior fixtures and furnishings from fire-withstanding aspects were related to the fact that the occurrence had not produced any tombstones. The accident occurred in Atlanta, Georgia, the largest airline hub in the southeastern United States. You know it has been said that when we die, we may not know whether we're bound for heaven or hell, but either way, we'll have to change planes in Atlanta. While the wreckage was in the

Safety Board's custody, an "open plane" approach was instigated to encourage other flight crewmembers to view the fire damage to the interior of the airplane. It was our hope the flightcrews would recognize the necessity for getting an airplane on the ground as soon as practicable after experiencing an in-flight fire.

The damage to the interior furnishings precluded a precise determination as to the cause of the fire. The assembled experts concluded it resulted from an electrical short circuit.

Some readers might remember the Canadian McDonnell-Douglas DC-9 flight that made an emergency landing at the Greater Cincinnati Airport after experiencing an in-flight fire. The landing occurred 17 minutes after the fire was discovered. At the time of the landing, visibility was virtually nonexistent in the cabin because of the smoke. The fire caused the loss of some flight control augmentation components and the Safety Board concluded the captain exhibited outstanding airmanship when he managed to execute a safe landing. A flashfire occurred in the cabin within 60 to 90 seconds after the doors and overwing exits were opened. Twenty three passengers failed to get out of the airplane and died in the fire. The Safety Board considered that it was possible some passengers became incapacitated before the landing due to exposure to the toxic gases and smoke. The Board further concluded that the cabin environment became nonsurvivable within 20 to 30 seconds after the flashfire. Toxicological studies showed the victims had elevated carbon monoxide levels ranging from 20 to 63 percent. The blood cyanide levels for the victims showed 0.8 to 5.12 readings. The Safety Board's report noted the incapacitating toxic levels for blood cyanide are between 0.5 and 0.7 readings.

After the accident, my phone started ringing, as the folks in Washington called to discuss the occurrence I had investigated in Atlanta. Thereafter, there was some action in the head sheds of both the FAA and Safety Board as they reviewed the regulations relating to the fire-resisting aspects of internal furnishings. However, my guess is that under similar circumstances today there wouldn't be much difference in the survivable aspects or the degree of damage to the interior of an airplane. So we probably haven't learned much from all this. But don't go away empty handed. If you're a regular airplane rider who never listens to the emergency briefings by the flight attendants, at least locate the nearest exit when you take your seat. If an emergency evacuation is ever necessary, proceed immediately to that exit and leave your carry-on luggage behind. You can't take it down the escape slides with you anyway, and the clutter of carry-on luggage around the emergency exits could prevent some passengers from surviving when a timely evacuation is essential.

PILOT MEDICAL CERTIFICATION

Let's look at a hypothetical situation that's not all that far removed from the real world. Say you're a senior airline captain. You have all the necessary aeronautical certificates and airplane ratings from the twin-engine McDonnell-Douglas DC-9 and Boeing 737 to the four-engine Boeing 747. Your seniority lets you choose your flights, and you make eight to ten round trips each month from New York to Los Angeles or over some other routes of equal or greater distance. Most of your neighbors are envious of your lifestyle because for not more than ten day's work each month you're drawing down a six figure salary.

Except for one little glitch or bugaboo, because all those certificates and ratings aren't worth a tinker's damn unless you also possess a first-class medical certificate. The certificates can be obtained only from an FAA-designated medical examiner. As an airline captain, you must renew your medical certificate every six months, and, if over 40 years of age, you have to show annually an absence of myocardial infarction on an electrocardiogram examination. As part of the renewal process, you have to make out an application whereon you certify to your past medical history and explain any significant changes since your previous examination.

You have an expensive home, a vacation home, a couple of late model cars, a nice boat, two kids in college, etc. It's time to renew that medical certificate, and lately you've been experiencing some chest pains, or some numbness on your left side, or some other physiological condition that is causing you concern. On your current application, would you list these changes since your last examination? No attempt will be made to provide an answer; however, many will admit this scenario has the potential to cause some applicants to fudge a bit by failing to provide some information that might be quite significant about the state of their health. Maybe that's why some pilots use designated medical examiners whose offices are located in cities distant from their domiciles.

- -

HOW DID THIS PILOT EVER OBTAIN A MEDICAL CERTIFICATE?

Dr. Joseph H. Davis, the Dade County, Florida Medical Examiner, asked the above question after reading the autopsy protocol for a fatally injured pilot. He was advised that it hadn't been acquired as the result of any cover-up by the pilot. Investigation showed he had provided a synopsis of his past medical history on each of his applications for medical certificates.

Dr. Davis is preeminent in his field, and credited with almost single-handedly writing the medical examiner legislation enacted by the State of Florida. He carried out the functions of his office with energy, gusto, and enthusiasm. Although the busiest man I ever met, he always responded favorably to our calls for assistance. Upon receiving the results of the post mortem studies that showed the pilot had an advanced cancerous condition that had spread significantly, I called Dr. Davis' office. After completing his review of the autopsy report, he praised the pathologist's performance and provided a list of several incapacitating eventualities that could have occurred.

The accident involved a six-place, twin-engine, corporate airplane that crashed in Mississippi while on a business flight. The pilot and five passengers were killed. Upon receiving initial notification of the occurrence while on standby duty at my Miami, Florida, residence, I was advised by the FAA inspector that the pilot had been employed by Eastern Airlines until about five years earlier, when he was denied a medical certificate. Before departing for the scene, I made the necessary telephone calls to ensure that the pilot's remains would not be embalmed until an autopsy was performed.

Clear weather prevailed, with excellent visibility. Witnesses observed the airplane dive into the ground while maneuvering to land at an uncontrolled (no control tower) airport. An 18-year-old youth at his family's dairy farm about 200 yards west of the crash site observed the airplane approaching from the southwest. When almost overhead, the plane banked sharply to the right and entered a dive. The witness yelled to his dad in the milking barn that the plane was going to crash. His father stepped outside in time to see it dive below the tree line. They ran to the scene intent on rendering assistance but were forced to back away when postcrash fire erupted. The cockpit and cabin were fire gutted. Examination of the wreckage of the almost new airplane showed no evidence of preimpact failure or malfunction of the aircraft structure, flight control system, or powerplant.

Dr. Davis had asked a pertinent question, and the thrust of the inquiry was directed towards the valid, second-class, medical certificate held by the pilot. Records showed he had had a tumor operation at the Walter Reed Hospital 17 years earlier that resulted in his removal from flying status and permanent grounding by the U. S. Air Force. Thereafter, he was issued a first-class medical certificate by an FAA medical examiner and found employment with Eastern Airlines as a pilot. He was forced to give up all flying seven years later when denied a medical certificate. The denial was related to the worsening effects of the cancerous condition in the sacral region, that had required five recent operations. His applications for medical certificates showed he had been medically discharged from the military and denied life insurance while employed by Eastern Airlines.

An airline captain is required to possess a first-class medical certificate. An airline copilot, corporate pilot, or a pilot engaged in other commercial flying for compensation, must possess at least a second-class medical certificate. A pilot who flies for personal reasons needs only a third-class certificate. However, third-class medical certificate holders are authorized to carry passengers, but not-for-hire. A first-class medical certificate is valid for six months for airline captain flying, one year for commercial flying and two years for personal flying. A second-class certificate is valid for one year for commercial flying and two years for personal use. A third-class certificate is valid for two years for personal flying.

The pilot made application for a third-class medical about two and one-half years after being denied the first-class certificate. The application for a certificate that would authorize him to engage in pleasure flying showed he had been self-employed in real estate for two years. The authorized medical examiner refused to issue him a certificate because of his preexisting cancerous condition. However, the pilot exercised his rights and appealed to the FAA. After interviewing the applicant, the assistant regional flight surgeon directed that he take a flight competency check with a designated pilot examiner, which the applicant passed without difficulty. Thereafter, he was issued a second-class medical certificate.

The probable cause of the accident was shown to be pilot incapacitation. However, it was my impression the FAA's assistant regional flight surgeon made a grave judgmental error. Any existing physiological condition that might lead to sudden incapacitation is reason for denial of any class of medical certificate. The pilot's aeronautical ability was not the issue. Flight competency checks are conducted for conditions such as loss of some fingers on a hand, or one leg shorter than the other, etc. The role of the flight examiner in such instances is to determine whether the particular physiological condition affects the applicant's ability to control the airplane and conduct other pilot duties.

The pilot's expectations were exceeded. Armed with his newly issued second-class medical certificate, he was authorized to exercise the privileges of his commercial flying certificate, and found himself another flying job. The medical examiner who denied his application for the third-class medical certificate said he didn't consider the pilot was medically qualified to perform any pilot duties whatsoever. While history has proved him right, that's little consolation to the next-of-kin of passengers who died in the crash.

- -

KNOWN MEDICAL DEFICIENCY

All three occupants perished during the crash of a small, twin-engine airplane being flown by a pilot with a known medical deficiency. The rather well qualified 57-year-old pilot was making a night instrument landing approach. The weather showed a 300-foot ceiling, two miles' visibility, with a light drizzle and fog. Persons not all that familiar with aviation need to understand that such weather can create a very stressful situation for many nonprofessional pilots. The pilot possessed a third-class medical certificate issued almost two years before the accident that was valid from a date standpoint, but invalid because of recent changes in his medical history.

The autopsy protocol showed severe coronary artery disease. At the time of the accident, anticoagulant and cardiovascular medication were prescribed for the pilot. A review of the pilot's medical history showed he had experienced myocardial infarctions, hereinafter called heart attacks, 13 years previously, with another occurring six months before the accident. The pilot's next-of-kin related he voluntarily grounded himself after the second attack. However, about a month before the accident, he did some flying with a flight instructor before commencing again to fly his airplane for business purposes. The two fatally injured passengers were company employees.

The examination of the wreckage showed nothing significant to the cause of the crash. For a comprehensive review of the pilot's medical background, assistance by the FAA's regional flight surgeon in Atlanta, Georgia, and the medical doctor assigned to the Safety Board's Washington, D.C. headquarters was requested. Their participation materially contributed to the Safety Board's probable cause determination that showed pilot incapacitation.

This accident graphically illustrates pilots' responsibilities insofar as medical certificates are concerned. In instances where a change occurs in pilots' physical well-being that might make them unable to meet the physical requirements of their current medical certificates, the regulations require that they obtain clearance from a designated aviation medical examiner before flying again.

- -

UNKNOWN MEDICAL DEFICIENCY

CASE NO. 1

I investigated several accidents caused by incapacitation where the physical deficiencies were unknown to both the pilots and their medical examiners. Six people died in one instance when a well-to-do pilot/rancher was executing a night visual approach to his private airfield in a sophisticated twin-engine airplane he owned. An active duty U. S. Air Force aviator occupied the copilot's seat. When I arrived at the scene, the sheriff said the pathologist wanted to see me. He was in the morgue doing the post-mortem studies on the victims. While I was standing on the opposite side of the table where he was working, he held something up, and after fingering it, I agreed with him that it felt hard as a nail. He then advised it was a section of the pilot's aorta and said no blood could have passed through it for years. He wanted to discuss what we were going to do about such a finding; however, in consideration of my weak stomach, I requested that we wait until he had finished his chores.

The autopsy protocol showed severe coronary arteriosclerosis, and scars in the heart indicated previous heart disease consistent with either myocardial infarctions or rheumatic fever. The pilot's next-of-kin were extremely cooperative and stated he never exhibited any abnormal physical difficulties. They reported he regularly visited the Mayo Clinic in Minnesota for physical checkups that included electrocardiogram examinations. Again, the assistance of the FAA's assistant regional flight surgeon in Atlanta, Georgia was requested. A comprehensive review of all the pilot's medical records showed no evidence that he suffered any cardiovascular difficulties. His doctor at the Mayo Clinic stated: "Review of his record does not disclose to me any predictable cause for sudden failure as a pilot."

It was learned it is possible for a person to live with an obstructed aorta. The medical folks informed me the tiny capillary blood vessels can expand as the main blood vessels become clogged. This provides an alternate route through which the blood can flow. That was the reason the pathologist was concerned. He said he had long been of the opinion that an electrocardiogram examination in conjunction with some sort of stress test would be far more beneficial than just the cardiograph examination itself. While the discussions were very interesting, they were focused on a field where my knowledge was sketchy at best. With assurances by the participating medical authorities that they would study the subject and take appropriate corrective measures, I returned to greener pastures insofar as my line of endeavor was concerned.

UNKNOWN MEDICAL DEFICIENCY, CASE NO. 2

Another investigation concerned an active duty U. S. Army aviator flying a civilian single-engine airplane for skydivers. The flight climbed to 6,000 feet where the skydivers jumped out of the airplane. Standard procedures stipulated the airplane would remain above the jumpers until all were on the ground. However, in this instance, the airplane went into a steep dive shortly after the skydivers jumped and impacted the ground at high velocity while the skydivers were still descending.

While documenting the accident scene, the medical examiner advised that his studies showed the pilot suffered a massive heart attack. Both coronary arteries were almost completely occluded by plaque deposits. The pathologist concluded the decedent suffered a massive myocardial infarction just prior to the crash.

In view of my first-hand knowledge of the strict medical standards employed by our military services, I was not surprised when the investigation showed the 37-year-old pilot had no history of cardiovascular disease. A few pilots with disqualifying medical deficiencies will probably continue to fall through the net. Keeping their numbers as low as possible is about the best we can ever hope for.

- -

GOODBY CRUEL WORLD

Two investigations involved pilots who dove their airplanes into the ground after leaving suicide notes. While feeling deep sympathy and compassion for the victims and their next-of-kin, it doesn't require much brainpower to conduct an adequate investigation. In another rather bizarre occurrence, a pilot took an airplane without proper authority. The airplane crashed after he killed himself with a revolver shot to the head. The real head scratching facet of that investigation related to the reasons why he also chose to destroy the airplane.

- -

ON THE AUTOMATIC PILOT?

Early one morning before leaving home for our office, notification was received that a sophisticated, twin-engine, turboprop-powered airplane had crashed in southern Georgia. The airplane was not on a flight plan and there was no record that the pilot communicated with any facility. None of the radio and navigation equipment frequencies found in the wreckage correlated with those in use in the vicinity of the crash site. Initially, we weren't able to identify the pilot, who showed a blood alcohol level of 0.27 percent. Upon moving the wreckage, we found the pilot's wallet that showed a New Orleans, Louisiana address and aeronautical charts showing that the radios and navigational equipment were tuned to frequencies in use in the vicinity of New Orleans.

Investigation disclosed the New Orleans-based airplane made a flight to Houston, Texas the evening before the crash. Upon arrival, the pilot requested that the airplane be refueled, and accompanied the passengers to some sort of meeting or gathering. He later returned to the airport and took off on the return flight to New Orleans. It was determined the fuel tanks were empty at the time of the crash.

The airplane manufacturer was requested to compute the point where the airplane would experience fuel exhaustion if it took off from Houston and continued on a direct course after overflying New Orleans. Not surprisingly, they came up with a location not far removed from the actual crash site. The pilot apparently engaged the automatic pilot after taking off from Houston. Thereafter, he either fell asleep or otherwise became incapacitated and the airplane continued on course until it ran out of fuel. Utilizing an individual's normal alcoholic metabolism for the approximate three-hour flight, we extrapolated back and determined the pilot's blood/alcohol concentration at takeoff was slightly more than 0.30 percent. And at that concentration, most folks are fairly well under the influence.

Another occurrence involved a well-qualified pilot on an instrument flight plan from Columbus, Mississippi to

Mobile, Alabama. When Memphis Center advised the pilot to contact Houston Center, there was no response nor were there any further communications with the flight. Controllers assumed a radio failure, and observed on their radar screens as the airplane overflew Mobile at the assigned altitude of 7,000 feet and continued out over the Gulf. A U. S. Naval airplane was being vectored on an intercept course. However, radar contact was lost at a point about 100 miles south of Mobile before the intercept could be effected.

Coast Guard search efforts were negative. Sometime later, an oil company vessel found a wingtip from the airplane in the Gulf; however, the pilot, and the remainder of the airplane were never found. A tower controller in Mobile reported the track and altitude of the airplane never varied after he accepted the radar handoff at a point 35 miles northwest of the airport to the limits of his radar coverage south of the airport. This indicates the automatic pilot was probably engaged. Obviously, the pertinent factors relating to the apparent incapacity of the pilot could not be determined.

- -

NOTEWORTHY PERFORMANCE

Just one more incident to close this discussion in a lighter vein. The student pilot, on his very first supervised solo cross-country effort, filed a visual flight plan and departed on the 110-mile trip to his destination. After takeoff, the tower controller advised the pilot to contact departure control. The controller provided radar vectors until the flight was clear of other air traffic, and then advised the pilot to resume his own navigation. While this is a fairly routine practice, it was something the student pilot had not expected, plus it interfered with the very careful and detailed preflight planning he had accomplished. The pilot became somewhat confused, and in the excitement that followed, began hyperventilating. Such rapid breathing over-oxygenates the blood, which can cause dizziness, fainting, and sometimes a feeling of stiffness in the joints.

The substance of the pilot's account of the mishap follows:

> **Feeling his ability to fly the airplane was impaired, he decided to execute a precautionary landing. He selected what appeared to be a suitable site, but upon leveling off, found he had overshot the field. Observing that the trees at the edge of the field were getting bigger and bigger, he did a "U" turn. However, before he could get the airplane down, the trees that had been behind him were now looming large ahead. Realizing he couldn't continue with the landing and avoid a collision with the trees, while noting increased dizziness and joint stiffness, he elected to force the airplane on the ground. After touchdown, he applied full rudder and the wheel brakes on one side. The airplane turned sharply and sustained substantial damage before sliding to a stop. In his concluding remarks, the pilot gave himself a pat on the back for the rather expert airmanship he exhibited while coping with his initial encounter with a bona fide emergency situation.**

HUP-TWO-THREE-FOUR

The Safety Board has jurisdiction in instances where military airplanes are involved in accidents with civilian models. Midair collisions were the norm, but whatever the nature of the accident, my policy was to extend an invitation for the military investigators to become parties to the Safety Board's investigation. Since there are open discussions of all the evidence, and a free flow of information between parties to an investigation, it was considered such an arrangement satisfied the requirements of both the Safety Board and the military. However, my initial joint investigation evolved into an example of well intentioned strategies coming unglued.

BREACH OF FAITH

A nonfatal accident involved a midair collision between a single-engine military jet and a small, privately owned civilian airplane. The military jet was descending to land in a "teardrop" pattern and the civilian airplane was in level cruise flight. The jet sustained relatively minor damage and landed safely. The civilian airplane lost an outboard segment from a wing, but the pilot executed a successful ditching just offshore in ocean waters.

At the initial meeting with military investigators, it was agreed a joint investigation would be conducted. There was no air traffic control involvement, and the collision occurred in what's called a "see and be seen environment." In lay terminology, that means the collision avoidance measures to be employed by the occupants of both airplanes were ever alert eyeballs mounted on near constantly swivelling heads.

The investigation appeared to progress in an orderly fashion. None of the occupants on either flight saw the other airplane before the collision occurred. The pilots and passengers on small airplanes are generally on the alert for converging traffic from ahead, the left, or the right. In this instance, the jet while in a steep dive descended into the top of a wingtip of the other airplane. Personally, I considered the military flightcrew was more at fault because they overtook the other airplane from above.

During our final meeting in a conference room on the military base, the senior military investigator requested that the statements obtained from the jet crewmembers be left for minor revisions. He promised the amended versions would be forwarded in several days. This presented no problems, and I readily agreed. Several days after returning to my office, a telephone call was received from the military representative. He said they had received instructions from higher headquarters that prohibited them from providing copies of their crewmember's statements to the Safety Board. He stated he would attempt to obtain other statements from their flightcrew to fulfill our requirements.

It was my feeling that they really "done me wrong." To say I was shocked would be a gross understatement. The accident occurred before I stopped erecting fences and started building bridges in my personal relationships. Without question, I gave a poor performance of concealing the fact that their actions had managed to elevate my dander to a new high. Since the conduct of the investigation was under the jurisdiction of the Safety Board, there was no such thing as higher military authority. We had no concerns about revised statements by their flightcrew; however, they were dead wrong when they refused to return the original statements obtained during the Safety Board's investigation.

Employing a rather firm tone, I informed the senior military investigator that the Safety Board would require no further assistance. Thereafter, Registered Mail-Return Receipt letters were mailed to both of the jet crewmembers. The cover letter requested their statements of the facts, circumstances, and conditions of the accident to the best of their belief and knowledge, as required by Safety Board's regulations. The correspondence included copies of the Safety Board's regulations regarding accident-reporting requirements by pilots and other crewmembers. A short time later, statements were received from both crewmembers of the military jet.

All of my subsequent assignments to investigate accidents involving military aircraft were conducted jointly. However, I delayed the onset of the proceedings until the military personnel stated that they fully understood that they were participating in investigations where the Safety Board had jurisdiction.

- -

MISUNDERSTANDING

A misunderstanding with military personnel surfaced during another investigation. The accident involved reserve officers who rented a single-engine civilian airplane for transportation to a training session. The mishap occurred while they were preparing to land at the destination airport on the military reservation. After experiencing an engine failure, the airplane crashed alongside the runway, killing the three officers on board. In accordance with our response procedures, I proceeded to the scene immediately after receiving notification of the accident. Upon arrival, I conducted the investigation with a great deal of assistance from the participating FAA inspector.

The primary thrust of the investigation was directed towards a determination of the factors relating to the engine failure. The airplane was equipped with two fuel tanks, one in each wing. One fuel tank was ruptured during the crash sequence, but the tank in the other wing appeared to be full of fuel. In most instances, experienced investigators are able to determine whether structural failure or flight control system malfunction was involved. Additionally, there is a reliable method of examining small reciprocating engines at the crash site that does not require complete disassembly. When nothing significant is found during such examinations, it is reasonable to assume the engines were capable of operation if provided with a combustible fuel/air mixture. In this instance, nothing abnormal was noted during the engine examination conducted at the crash site.

The one thing left to do was carefully drain and measure the fuel remaining in the airplane. The wreckage was not in a level attitude, and care was taken to assure that no fuel was spilled while draining the tank. The measurements confirmed that the tank was filled to capacity.

The evidence showed the entire flight had been conducted on fuel from the ruptured tank. In view of the elapsed time from takeoff to the accident, it was reasonable to conclude the engine failure was caused by fuel starvation. Under such circumstances, mismanagement of the fuel system by the flightcrew becomes the predominating factor in the accident. While we were collecting our equipment and expressing our appreciation to base personnel for their assistance, a military plane landed, taxied as close to the crash site as possible, and a team of military officers from another base deplaned. After the introductions, it was learned they had been dispatched to conduct an investigation of the accident.

The team consisted of a major and two captains. To say they were surprised to find we had disturbed the wreckage in any manner whatsoever would be putting it mildly. One of the captains was so vehement we felt they might summon the Base Provost Marshal for assistance in dealing with what he considered our "too hasty and unauthorized" tampering with the wreckage.

However, the cooler heads prevailed, and after the situation quieted down, the military team was invited to become a party to the Safety Board's investigation. Such an arrangement would give them access to all the evidence we had collected, including prints of our photographs. They were also advised that the Safety Board probably had jurisdiction, but positive clarification of the issue would be obtained before the day was over. As it turned out, the Safety Board did have jurisdiction, and we conducted the remainder of our on-scene activities, and necessary future liaison, in a friendly and cooperative manner.

It was relatively simple to determine the probable cause of the accident. But, in retrospect, suppose the tank found full of fuel had ruptured during the crash sequence. If that had been the case, we would probably still be advancing theories about how the accident could have happened.

- -

NOT QUITE BY THE BOOK

The occurrence involved a small single-engine airplane belonging to a military aero club. Such aircraft have civilian registrations, and the Safety Board has jurisdiction when they are involved in accidents. The airplane was flown by an experienced, active-duty, jet fighter pilot. The accident happened while he was attempting to return to the airport after experiencing a loss of engine oil pressure shortly after takeoff. Both the pilot and his daughter were killed in the tragic mishap.

The accident occurred on the first flight in the airplane following a mandatory periodic inspection. The pilot had pushed the airplane out of the hangar, accomplished the preflight inspection, started the engine, and taxied to the active runway without outside assistance. While holding in the area short of the runway, he accomplished the usual engine runup and preflight checks before receiving takeoff clearance from the tower.

The flight climbed out on the runway heading, but about two minutes after takeoff, the pilot radioed he was returning because of a loss of oil pressure. Upon executing the turn back to the airport, the flight was over a large unobstructed grass field that would have provided an excellent emergency landing site. While the airplane was returning to the airport, the engine failed completely. The pilot attempted to turn back to the open field that was now behind him. However, he lost control for some reason, and the airplane crashed in woods at the edge of the field.

A significant oil spill was observed on the ramp at the point where the engine was started. Thereafter, it was possible to follow the exact taxi path of the airplane by reference to a trail of oil on the pavement. At the point short of the active runway where the pretakeoff checks were conducted, another significant oil spill was observed.

Postcrash fire occurred and the engine compartment, cockpit, and cabin were fire gutted. The fire destroyed a great deal of significant evidence. Review of the aircraft logs and records showed the oil and oil filter were changed during the periodic inspection preceding the fatal flight. FAA regulations stipulated a test run of the engine was required before the airplane could be returned to service. The records failed to show that the mandatory engine runup had been conducted.

When the inspector who signed off on the records was queried about the matter, he stated an engine runup had not been accomplished. He was astonished, and in view of the consequences of the accident, extremely dejected, when the FAA regulations were opened to the section showing a test run of the engine was a requirement before he was authorized to return the airplane to service. The inspector had been employed by the aero club since retiring from active military service six years earlier. He readily admitted he had never known engine runups were required in such instances.

The evidence showed the oil supply slowly leaked out of the engine from the time it was started. The only accomplishment during the annual inspection that might have been related to the oil spillage was replacement of the oil filter. The oil filter components sustained significant postcrash fire damage. However, it was reasonable to assume the filter was not properly tightened against the gasket, or the gasket crimped during installation, leaving a recess through which the oil could spill overboard.

Complete loss of oil pressure is never something to be taken lightly in airplanes, or elsewhere. The pressure gauges might have failed; however, in instances where the oil and cylinder head temperature gauge readings are rising in conjunction with a complete loss of oil pressure, there is reason for grave concern. No attempt is being made to suggest pilots always execute an immediate forced landing solely upon indications of a complete loss of oil pressure. However, depending upon the proximity to an airport, in conjunction with terrain features, a precautionary landing that might result in some damage to the airplane may well be the most appropriate course of action in some instances.

A complete loss of the engine oil supply is not a common happening in aviation. My investigations included seven oil depletion occurrences during a 20-year tenure with the Safety Board. In cases where operation of the engine continued after the loss of all engine oil, it was always found "seized," meaning rotation of the crankshaft was not possible. Disassembly inspection of "seized" engines showed the internal components had been subjected to extreme overheat distress that usually resulted in the sides of the pistons being almost "welded" to the cylinder walls. During discussions with engine manufacturer's representatives, FAA maintenance inspectors, and certificated aircraft mechanics, it was learned internal combustion aircraft engines will normally seize about two minutes after complete depletion of the oil supply.

It was difficult to fathom the loss of control by the highly experienced pilot. Investigation showed he had flown the particular airplane model on only one other occasion since joining the aero club. The retractable landing gear was found extended in the wreckage. Under the circumstances of the accident, this provided the basis for some head scratching.

The landing gear system incorporated a unique feature that provided automatic extension whenever the airspeed decreased to a predetermined level. The automatic extension feature could be disabled through use of an override button in the cockpit. It was theorized the pilot may have failed to hold the landing gear in the retracted position with the override button. In such an event, sudden extension of the landing gear while in the steep turn back to the open field could have contributed to the loss of control.

Aviation accident investigators occasionally find themselves second-guessing the actions of others while sitting around pondering the reasons for accidents. Without any finger pointing, we did some meditating on the landing gear features of the airplane. Obviously, the manufacturer provided for automatic landing gear extension to prevent inadvertent wheels-up landings. While we didn't come up with anything concrete, it was concluded engineers have to exercise extreme caution when attempting to provide solutions for one problem area, to assure they don't open other avenues that can lead to calamitous consequences.

WORRIED MIND

A highly decorated WW II fighter pilot was killed while executing a night approach in visual weather to a joint-use military/municipal airport. The airplane crashed into trees short of the runway after the pilot advised the tower controller he intended to land near the end of the runway to permit a turnoff at the first taxiway leading to the aero club ramp on the military side of the field.

Using Safety Board terminology, this was a "high visibility accident" that received wide media attention, especially on the local scene. The pilot apparently had lots of "PI" (political influence) and we received several inquiries from a senator's office in Washington, D. C. while the investigation was on-going. The evidence indicated the pilot had inadvertently flown his airplane into the treetops while attempting to land short in order to make the planned turnoff on the first taxiway.

Military personnel were extremely cooperative and provided much assistance during the conduct of the investigation. The airplane was owned by the pilot. We received a statement from the sergeant who managed the aero club, showing he helped the pilot move his airplane out of the hangar on the morning of the accident. Before taking off, it was refueled at the aero club pump.

Although there was no military involvement in the accident, it was noticed that a colonel from the Base Safety Office was very much in presence while we were collecting the evidence. I mentioned to him it would take several days to complete our comprehensive investigative efforts and, although he was welcome to participate if he desired, I'd be happy to come by his office and give him a briefing when the investigation was concluded.

He mentioned there was a matter of interest to the Commanding General. It related to the fact that the airplane was normally parked and serviced at the aero club facility. They were concerned we might expound on this subject in our investigative report. The colonel was advised that we were required to determine whether the airplane had been properly maintained and serviced, including a determination that there was adequate fuel on board to conduct the flight. However, unless there was some possibility that servicing and refueling of the airplane might be involved in the accident, the Safety Board had no interest in the locations where such activities were conducted. The expression on the colonel's face, and his subsequent remarks, showed I was profiting from my recent study of Dale Carnegie's book about winning friends and influencing people.

BEWILDERED

Feel frustrated, hearing strange words and phrases, hanging out with a bunch of strangers? Well, don't panic, it may only be that you are a jury member in a liability lawsuit resulting from an aircraft accident. When I was employed by the Safety Board in 1966, liability lawsuits were not too prevalent. However, at the time of my retirement in 1988, the numbers had increased dramatically. Near the end of my tenure with the Safety Board, investigators returning from the field phase of an inquiry sometimes not too jokingly remarked: "My response time was outstanding; I beat the lawyers to the accident scene."

Often, upon filing their cases, the plaintiff's sides employ the shotgun approach when naming defendants. They routinely sue the government, airframe and component manufacturers, maintenance facilities, individual repairmen or inspectors, and anyone else who might have been involved with the airplane, or the conduct of the particular flight on which the accident occurred. This shotgun approach method of naming defendants can be expected to continue until such time as laws are passed that require the plaintiff's side to pay a defendant's legal fees if the plaintiff fails to prevail. However, with a Congress comprised of an overwhelming majority of lawyers, the probability for passage of such a bill is about on a par with the chances they are going to lower our taxes.

When hauled into court in a liability case, one of the defendant's basic rights relates to trial by an impartial jury. In most liability cases, lawyers representing the plaintiff's side have interpreted "impartial" to mean those with no knowledge or experience whatsoever about the subject matter involved. Accordingly, they resort to any means possible to exclude persons employed in the aviation industry and others with aeronautical knowledge and experience from serving on the juries.

While these practices do not necessarily result in a panel of unintelligent jurors, it does mean we have jurors in aircraft accident liability cases with, at best, only sketchy knowledge of aviation matters. The defense attorneys are usually representing the government, airframe or component manufacturers, repair facilities, etc. Routinely, their clients have a wealth of aviation knowledge and many experts available to render assistance. The plaintiff's attorney normally represents those injured or killed in an accident. In many instances, their witnesses do not have impressive aviation backgrounds.

Having a know-nothing jury from an aviation standpoint at the outset of a trial puts the plaintiff's side on equal footing with the defense. In other words, both sides start from scratch with the jury, and whichever side does the better job of educating, or, as the case may be, misleading or possibly confusing the jury, wins. And if it happens to be the wrong side, that's the way it is. The judges aren't to blame. They are present to ensure that the proceedings are conducted properly while making other rulings, but it is not their function to ascertain that the evidence presented is adequate, factual, or for that matter, truthful.

Some aviation accident juries arrive at, well, just let it be said defendants are being found liable in an inordinate number of cases. Please understand, this is not a tirade against our justice system. It is recognized that the circumstances of some accidents provide the basis for legal actions. And however flawed the present system of jury selection may be, it is light years ahead of turning the cases over to a panel of lawyers or experts.

In the event there was one aviation expert on the jury, it would probably be necessary to have a whole panel of experts in order to keep matters completely impartial. Otherwise, that individual might be able to lead the other jurors down any path he desired. In the event the opposing lawyers were permitted to settle the cases, they would slice up the pie in a manner that would fill all their pockets with greenbacks, and go their merry way.

The present system of jury selection has some desirable features. It would be a rare occasion to have a panel of six or twelve jurors without one or several members having above average intelligence. The intelligent jurors will probably be able to fully understand, and evaluate the evidence. Then, during the deliberations, they are in position to educate the other members. The panels usually arrive at a reasonable understanding of the issues involved. Accordingly, the present system is probably about as foolproof as any alternative arrangements that might be devised.

Reviewing the events that occur in a number of fatal accidents should shed some light on the subject. Let us say a noninstrument-rated pilot takes off in instrument weather conditions, or, as the case may be, continues a visual flight into instrument weather. After becoming disoriented or confused as to his or her location, the pilot uses the radio to call for help. An air traffic controller receiving the transmissions answers. In the event the airplane subsequently crashes, the government often loses, regardless.

In reality, the controllers, and ultimately the government, find themselves in a near no-win situation. In the event a controller, after determining the pilot is not instrument rated, refuses to render assistance, the government might be charged with neglect. If assistance is provided, the government often loses because the plaintiff's side claims the controller was providing vectors, or other instructions, to the noninstrument-rated pilot in instrument weather conditions. And since our country is already operating in the red, there is no way out for us taxpayers, the ultimate source of that green-shaded paper the government uses to pay its bills.

Understand, in all such cases the pilots are at fault. If they initiate a visual flight in instrument weather conditions, they have violated the FAA's regulations. Likewise, if they continue a visual flight into instrument weather conditions, they are again violating FAA regulations. According to the regulations, a pilot who encounters instrument weather while on a visual flight "shall remain VFR (visual) and land." However, instrument-rated pilots do have the option of remaining visual while requesting an instrument flight plan. When their flight plans are processed and approved by air traffic controllers, they are authorized to continue to their destinations.

Another area in which the government often loses involves the preflight weather briefings provided pilots. Looking over some jury verdicts might lead us to believe weather forecasting is an exact science. But everyone knows that is not true. Without being disparaging, we didn't sometimes refer to forecasters and briefers as "weather guessers" without reason. Basically, the forecasting system is founded upon what happened when similar meteorological conditions or weather patterns existed in the past. Using modern technology, the forecasts are improving, but they will never be completely accurate.

During my tenure with the Safety Board, I gave many depositions relating to accidents where absolutely no one could have been at fault except the pilot. Regulations precluded me from discussing the cases with the attorneys before giving my testimony. However, that restriction was lifted after the deposition proceedings were concluded. A plaintiff's attorney was once queried as to how they ever arrived at the basis for a claim against anyone as a result of the particular accident in question. He replied: "They pay off like clockwork; we don't win them all, but we win enough to make it worth-while."

On another occasion, an attorney, who happened to be a friend of mine, agreed to an out-of-court settlement. I asked why he hadn't persisted with the case, explaining all he had to do was bring a particular aircraft component into the court, educate the jurors as to just how it functioned, and he would never have been subjected to an unfavorable verdict. He replied: "We settled for less than would have been required to pursue the matter through the courts."

After my retirement, another career as a consultant in aviation accident lawsuits was contemplated. Primarily, I would have to be retained by the defense. I could never compromise myself by working or testifying as a "hired gun." That's the moniker we attach to someone willing to give testimony designed to blame accidents on anybody or anything other than the actual factors involved. I was retained as a consultant on several occasions, but never got as far as the witness stand. Being deprived of the opportunity to testify in court prevented me from obtaining a track record. Accordingly, my anticipated career as a consultant never got off the ground.

In every instance where I was involved, the opposing sides reached an out-of-court settlement. Upon asking our attorneys why they agreed to give the plaintiff anything, they responded: "We settled for less than would have been required to pursue the case through the courts." On several occasions, they said the opposition probably agreed to the settlement because of testimony I was expected to give.

My involvement in the liability lawsuits provided insight into matters I wasn't previously aware of, and something the general public probably isn't aware of either. We may be misinformed from watching so many "who-done-it" TV shows where, about five minutes before the final commercial, the attorneys for one side bring forth some surprising bit of evidence that lets them instantly win the case. Usually, the surprise evidence relates to matters the opposition didn't know, resulting in their being caught completely off guard. At any rate, it was learned that is not the real world in liability cases arising out of aircraft accidents.

The opposing attorneys passed back and forth interrogatories showing precisely the testimony they expected to elicit from each witness they intended to call. Additionally, they provided the other side with a dossier on each witness. Obviously, these actions provided the opposing attorneys with enough information to make a determination as to the weight or significance to attach to the expected testimony by the witnesses. From my viewpoint, these doings constituted very much of a buddy-buddy arrangement between the opposing attorneys.

A high percentage of accidents in which lawyers are injured or killed generate liability lawsuits. Next in line are professional people, followed by those high up the corporate or business ladder. In other words, the more money victims were earning, the greater the chances a lawsuit will be filed in their behalf.

The defense called me as a consultant on a case involving a lawyer-pilot flying a light, twin-engine airplane that crashed after both engines failed. The fuel valves in the wreckage showed both engines were positioned to the left wing tank. Fuel consumption calculations showed the accident occurred after an elapsed time when fuel exhaustion would have been expected if both engines had been operated on the same tank throughout the flight.

The left wing tank was violently torn apart during the crash sequence; however, there was evidence showing the tank was empty at the time of the crash. It was not possible to determine precisely the quantity of fuel in the right wing at the time of the accident because of ruptures to the wing tank during the crash sequence. However, a significant quantity of fuel was found in the right wing tank during the investigation. The wreckage had come to rest in a hayfield. When I visited the crash site some two years after the accident, it was observed that the hay was still not growing in the area where the right wing had come to rest because so much of the fuel had seeped into the ground.

The evidence indicated the pilot had mismanaged the fuel system. It might even be said the accident happened just the way he programmed it, albeit it was unintentional. The plaintiff's side hired a consultant with an impressive background. However, he did have a very formidable task. First, he would have to convince the jury the failures of both engines were not related to fuel starvation as indicated by the evidence. Secondly, he would have to come up with separate events that had caused both engines to fail almost simultaneously. Mission impossible insofar as aircraft accident investigators are conerned. However, the presentation of such flimsy testimony sometimes results in a favorable verdict for the plaintiff's side.

The defense attorneys who retained me forwarded all the interrogatories plus the dossiers on the plaintiff's witnesses. It was learned their consultant intended to give testimony that his examination of the left wing tank to which the cockpit fuel valves were positioned showed deformation from hydraulic forces. Apparently, he hoped this would indicate to the jury there was sufficient fuel remaining in the tank for it to slosh forward during the crash sequence with enough force to cause the forward end of the tank to expand, or take on a balloon-like deformation. Next, he intended to testify how a vapor lock in the fuel lines to one engine, and water contamination of the fuel supply to the other engine, had caused the engine failures. He's probably lucky the case never came to trial, because it is doubtful he could have sold those theories to any jury. Incidentally, I was primed to refute his testimony. However, almost at the courthouse steps, the opposing sides reached an out-of-court settlement.

Law firms that might be categorized as "ambulance chasers" have become more and more involved on the aviation liability scene. The journeyman investigator has little difficulty spotting evidence that might provide the basis for yet another lawsuit. In one occurrence, a single engine airplane crashed into trees off the left side of the runway while taking off at a small uncontrolled airport. We collected statements from aviation orientated witnesses showing that the airplane lifted off prematurely, and remained airborne in a semi-stalled, wings rocking, nose high attitude, until it drifted into the trees off the side of the runway.

The wreckage examination is carried out to determine if there is evidence that a failure or malfunction of the aircraft structure, flight control system, or powerplant was involved in the accident. In this instance, the structure and powerplant examinations showed nothing significant. However, the examination of the flight control system showed a fracture in the metal on the elevator control bell crank in the tail. The area in the vicinity of the fracture showed bending deformation. Examination of the fracture surface showed no evidence of fatigue or previous cracks.

Some airplanes are equipped with rigid rod, push-pull, flight control systems wherein the components are subjected to both tension and compression forces. The accident airplane was equipped with a cable flight control system. Since there are no compression stresses in a cable system, the fractured bell crank could only be subjected to tension forces during normal operations.

Evidence indicated the fractured bell crank was not involved in the accident. It was obvious it had occurred during the crash sequence because a piece of metal being subjected to tension stresses only during normal operations will not show bending deformation. The participating FAA inspector agreed with me that the fractured bell crank had absolutely nothing to do with the accident.

Realizing it would be necessary to explain the fracture in my narrative report, a comment was made to the FAA inspector about the telephone calls I would be receiving from experts hired by attorneys who had combed my factual report for evidence that might provide the basis for a liability lawsuit. My prediction proved to be accurate. And whether authorized or not, I provided them with the above analysis when they called. Still, several made arrangements to examine the wreckage that had long since been released from the Safety Board's custody.

When questionable evidence was uncovered, Safety Board investigators and FAA inspectors often joked with the airframe and component manufacturer's representatives about their upcoming lawsuits. In one occurrence, an expensive, high performance, single-engine airplane sustained an in-flight structural breakup after it came screaming out of a thunderstorm. The pilot and four passengers were killed. The noninstrument-rated pilot, who was suffering from a severe case of get-home-itis, received a preflight weather briefing that showed marginal visual weather conditions prevailing over his planned route of flight, with the probability of significant thunderstorm activity. The weather briefer advised the pilot that visual flight was not recommended. Visual weather conditions prevailed at the departure airport, and there were no regulations that prevented the pilot from initiating the flight. However, once airborne, the pilot was required to comply with the rules for visual flight that precluded him from penetrating the weather conditions he encountered while en route.

Examination of the fracture surfaces on the components that separated in flight showed no evidence of previous cracks or fatigue. Occurrences involving in-flight breakups always required lengthy investigations. In this instance, the debris had to be gathered from a half-mile-long scatter pattern to make a two-dimensional layout. A study of the bending deformation of the separated wing and tail components showed the structural breakup resulted from positive aerodynamic stresses in excess of the design limitations. Such overloads are invariably induced by the pilots as they manipulate the flight controls while attempting to regain control.

At the time of the accident, some defensive tactics had crept into my investigations. Meaning, since we had learned what the lawyers or their experts were looking for when they combed our reports, more time was being spent documenting the wreckage. An FAA inspector, and representatives from both the airplane and engine manufacturer, were participating in the investigation. Although an intensive search was conducted, we were unable to locate one of the elevator counterweights. Counterweights are attached to the ends of the ailerons on the wing, and the rudder and elevators on the tail, to inhibit flutter, and otherwise stabilize and balance the control surfaces in the wind stream.

Efforts were always made to locate the flight control counterweights for accidents involving an in-flight structural breakup. However, finding all the counterweights can be an arduous if not impossible task. In this instance, the search efforts were hampered by wooded, brush-covered, soft terrain, and a river flowing across the wreckage path.

In view of other evidence collected during the investigation, we knew the missing elevator counterweight had no bearing on the accident. However, while unwinding over a cup of coffee at the conclusion of the on-scene investigation, the FAA inspector and I were kidding the aircraft manufacturer's representative about the difficulty they would encounter in defending themselves from the claims of those ominous liability attorneys. We discussed how all they would have to do was have some "hired-gun" explain to the jury how separation of the elevator counterweight had resulted in an unbalanced condition in the tail that caused it to swish back and forth until it finally broke off, and the jury would surely render a verdict in favor of the plaintiff's side.

The aircraft manufacturer's representative interrupted our carryings-on to say he wasn't through looking for the missing counterweight. Before we departed from the scene, he offered a reward to anyone who found it. About a month after the accident, he called and reported the partially buried counterweight had been found in the wreckage scatter path.

Let's review a couple of aircraft liability lawsuits where the outcomes should raise some eyebrows as to the fairness of judgments rendered against manufacturers. A sophisticated, high performance, twin-engine airplane, equipped with internal combustion engines, crashed when both engines failed shortly after takeoff. Investigation revealed it had been serviced with jet fuel prior to the flight. The plaintiff's side received a multi-million dollar settlement that ultimately went against the aircraft manufacturer because the other defendants did not have sufficient assets or insurance to satisfy the judgment.

In another case, a pilot possessing no aeronautical certificates or ratings whatsoever, crashed in an airplane he had stolen. Here again, the plaintiff's side won a multi-million dollar settlement that ultimately went against the aircraft manufacturer.

Maybe we have too many legal-beagles. The United States, with five to six percent of the global population, is the place of employment for approximately two-thirds of the world's lawyers. On top of that, about 90 percent of all product liability lawsuits are filed in our country.

In 1991, lawyers in Florida became subject to stiffer advertising rules and more detailed financial disclosures in their efforts to find liability claimant clients. The day after the rules took effect, some frustrated lawyers sued, claiming the rules were unconstitutional, vague, and unfair. The Florida Bar and all the State's Supreme Court Justices were named defendants. The lawsuit shouldn't come as any great surprise when considered in light of the above-mentioned attorney's comment about the results of liability lawsuits: Remember, he said: "They pay off like clockwork; we don't win them all, but we win enough to make it worth while."

All these legal maneuverings in conjunction with excessive government regulation have contributed to a significant wind-down in our light airplane manufacturing industry. Increased production costs mandated by government regulators from a host of departments or agencies, and liability judgments against the manufacturers, have to be passed on to the aircraft purchasers. The Piper Corporation is in bankrupsy proceedings, and there has been some slowdown at other light airplane manufacturing facilities. In a manner of speaking, they are killing the gooses that were laying lots of golden eggs. Light airplanes are needed on a worldwide basis. Like so many other products, the factories are springing up in foreign countries.

AMERICA'S KAMIKAZE CORPS

Unlike their Japanese counterparts who could only obtain the desired objective by crashing their bomb-laden airplanes into enemy warships, ours have devised multiple methods of saying good-by to this old world. Some of the more prevalent means they employ involve flights in airplanes that have not received proper maintenance and inspections; attempting flights with a known discrepancy in some aircraft system or component; initiating or continuing flights with a questionable supply of fuel; attempting to skim across the terrain at the lowest possible height as part of a thrill-seeking adventure; and other measures not designed to contribute to the attainment of a ripe old age. Accidents resulting from the above-mentioned factors are not rare occurrences. Collectively, they accounted for about 20 percent of 679 investigations the author conducted during his 20-year tenure with the Safety Board.

Admittedly, our Kamikaze pilots did not take off with the conviction they had been given this wonderful opportunity to obtain eternal glory by dying for their Emperor, or some other cause they may have considered just. So, in the true sense of the word, ours were not really Kamikaze pilots. However, they managed to kill themselves on more occasions than most folks would ever believe. I recently read statistics showing Japan's Kamikaze pilots flew 1,228 suicide missions in WW-II during which they sank 34 ships and damaged 288 others. In the process, they inflicted in excess of 15,000 casualties, including more than 5,000 killed. While these figures might sound impressive, a tabulation of all the aircraft accidents caused by the factors mentioned above would show that our "Kamikaze" pilots have destroyed a greater number of airplanes than the Japanese did, while inflicting even more casualties upon ourselves.

Insofar as serviceability is concerned, today's light general aviation airplanes are extremely reliable. Likewise, the reciprocating engines currently installed in small airplanes have proven to be almost trouble-free. With proper care and maintenance, modern-day airplanes and engines are built to give satisfactory service over an extended period of time.

The key words in the above paragraph are: "With proper care and maintenance." Otherwise, airplanes fall into the same throw-away category as automobiles, appliances, television sets, beer cans, etc. Pilots who operate airplanes that have not had the required maintenance and inspections are taking unnecessary risks. Others, who knowingly fly airplanes with a defect in some system or component, are jeopardizing their and their passengers' lives. At the very minimum, pilots who attempt such flights should realize they are engaging in a form of Russian roulette. Admittedly, the odds for successful completion of the flights are better than the prevailing one to six for the revolver. However, as shown by the following occurrences, such flights terminate in disastrous consequences on many occasions.

- -

SAGE ADVICE THAT WENT UNHEEDED

An Air Force sergeant on his way to the mess tent in a base training area on a military reservation stated:

> **About 10:30 p.m., I saw a flash in the sky. I thought it was a daylight flare. Then I saw something falling away from the fire, followed by the high pitched sound of an engine revving up. I then realized it was an airplane going down. An explosion occurred when the airplane impacted the ground and a fire glowed over the crash site.**

The pilot and his two passengers were killed. The accident involved a small, privately-owned, high-performance, pressurized, twin-engine airplane equipped with turbo-charged engines and three-bladed propellers. The airplane sustained in-flight separation of the left engine and about half of the left wing. The separated components came to rest several hundred feet from the main wreckage location.

Examination of the left engine showed it had been subjected to an intense in-flight fire. The fire was in the accessory section on the rear of the engine. The molten status of the fracture point on the left wing spar showed the wing had literally burned off while the flight was still airborne.

After examining the separated left wing and engine, I turned to the base security representative and told him the detailed examination of the wreckage would have to be delayed until we assembled a team with the requisite technical qualifications and expertise. Since the crash site was in a remote area on the military reservation, my request presented no problems. After requesting assistance by the experts, the participating FAA inspector and myself busied ourselves with other activities essential to any investigation while awaiting the technicians' arrival.

The pilot was returning to his home base on an instrument flight plan. The takeoff was made at 10:25 p.m. from a civilian airport on Florida's gulf coast several miles south of Eglin Air Force Base. The controller had cleared the flight to climb to his assigned altitude on a magnetic heading of 020 degrees. Two minutes after taking off, the pilot reported he had experienced an engine failure. The controller advised the pilot that he was less than a mile south of Eglin Air Force Base and cleared him to land on runway 12. The accident occurred five minutes after the takeoff while the pilot was maneuvering to land.

The pilot was experiencing difficulties with the left engine prior to his takeoff. At approximately 10:00 p.m. on the accident date, he telephoned the principal officer of a maintenance facility that had accomplished some recent repairs on the airplane. After discussing the problems the pilot was experiencing, the maintenance official said he advised the pilot to "have a mechanic check it out." Obviously, the advice was unheeded because a mechanic was not immediately available, and the flight took off about 25 minutes later.

A security guard on duty at the departure airport said he had noticed some "missing and sputtering" when the engines were started. He was not particularly observant of the events that transpired after the engines were shut down. When the engines were restarted a short time later, he again heard some sputtering. However, on the ensuing takeoff, he noticed nothing unusual in the operation of the engines.

The thrust of the investigation was directed towards the fire in the left engine's accessory section that caused the crash. The portion of the wing spars aft of the engine installations are protected by steel firewalls. Examination of the separated left engine and wing showed the intense fire in the accessory section had burned through the top of the aluminum wing skin forward of the engine firewall. Thereafter, the fire burned back down through the top wing skin just aft of the wing spar. The evidence was sufficient to substantiate the fact that in-flight "burn off" of the left engine and wing had occurred in this instance.

The left engine was removed to a nearby certificated repair facility. Many hoses and most of the electrical wiring in the accessory section had burned free of the attach fittings. It was readily apparent that significant evidence was destroyed by the fire. Insofar as possible, everything in the accessory section was accurately documented.

However, we were unable to determine the precise source of the in-flight fire. Laboratory analysis of debris and components from the accessory section proved to be fruitless. The only satisfaction obtained from our efforts was the realization that no significant evidence was destroyed during the investigation. Such is the life of an investigator. The circumstances of some accidents make it impossible to come up with all the answers.

But that did not stop the theorizing that was in abundance. In view of the circumstances, we were of the opinion that the fire had to be fuel-fed. Additionally, to generate sufficient heat to burn the wing off so rapidly, we were confident there had to be a ready source of fuel as opposed to the burning of some residual fuel that might have leaked into the engine accessory section. New hoses were fabricated and installed during the recent repairs and we considered it possible the fire was fed by fuel from a fuel hose that had either ruptured, or come loose at a fitting.

A pilot experiencing an uncontrollable wing fire is confronted with a dire situation. In such an event, the following actions are recommended:

> **IN THE EVENT OF AN UNCONTROLLED WING FIRE, DECREASE ENGINE POWER AND SLOW THE AIRSPEED. THEREAFTER, SIDE-SLIP THE AIRPLANE INTO THE NONBURNING WING IN ORDER TO ACHIEVE A RAPID LOSS OF ALTITUDE, AND LAND AS SOON AS POSSIBLE.**

That doesn't mean at the nearest airport or suitable emergency landing site. It means the airplane must be landed posthaste, pronto, right now. The only limiting criteria is the pilot should not endanger the lives of anyone outside the airplane. It is far less life-threatening to stall an airplane into the treetops, lake, ocean, or whatever else might lie below, than it is to have it dive into the terrain at high speed in an uncontrolled attitude after the wing burns off.

- -

There is the story about a very religious man who climbed on his roof during a flood. He refused to be rescued by a boat that came by, and, after the water kept rising, by a helicopter that hovered overhead. In both instances he told the rescuers: "The Lord will save me." However, the water kept rising and he ultimately drowned. Upon passing through the pearly gates, he asked why they hadn't done something to save him. St. Peter replied: "We sent a boat that you refused to be rescued by. Thereafter, we dispatched a helicopter in your behalf that you also refused to climb aboard. So, welcome to paradise, pick up your halo at the supply desk. Report to the hangar at noon to learn our air traffic procedures. They will issue a can of fleecy white WD-40 you are to use to keep your wings from corroding

THAT'S NOTHING TO WORRY ABOUT, IT'S BEEN LEAKING A LONG TIME

The airplane with the pilot and one passenger on board departed from the runway adjacent to the pilot's residence on a 45 minute flight to the Florida Keys. The retired airline captain they visited also resided alongside a private airstrip that provided ready access from his property to the runway. After landing, the pilot taxied the single-engine airplane right into his friend's back yard. Several hours later, they reboarded the airplane for the return flight. The statement by the retired captain they were visiting included the following:

> **"While observing the pilot starting the engine, it started momentarily and then stopped. I noticed a scraping or tapping sound that seemed to come from the propeller spinner. I mentioned it to the pilot and he said it might be a loose washer, or something to that effect.**
>
> **I had noticed when he arrived a large streak of oil on the left side of the engine cowling and mentioned it to him. He said it was normal, and had been occurring for a long time.**

The airplane was piloted by a retired airline captain with more than 27,000 flight hours. After taking off for the return flight, the airplane proceeded several miles east before turning to a westerly heading. When abeam of the airport, witnesses heard an explosive-like sound followed by a greater noise intensity as the engine and propeller speeds increased. Some of the witnesses observed an object fall from the airplane as it entered a steep left bank and descended into the terrain.

The pilot was killed and the passenger sustained serious injuries. The object the witnesses observed fall from the airplane proved to be the propeller spinner. The dome-shaped spinner fits over the propeller hub to provide streamlining. It was found resting on its nose about one-half mile east of the wreckage. The propeller cylinder, some cylinder attaching hardware debris, and several ounces of oil were found inside the spinner.

Many of the lighter, single-engine airplanes are equipped with fixed-pitch propellers that have no propeller controls in the cockpit. The speed at which a fixed-pitch propeller rotates is determined solely by the throttle setting selected by the pilot. The accident airplane was equipped with a constant speed propeller. In such configurations, the pilot obtains the desired propeller speed through use of a control lever that is normally located alongside the throttle in the cockpit. Basically, movement of the control lever signals the propeller governor to vary the pitch (load) on the propeller, as required, to maintain the desired revolutions per minute.

The propeller governor utilizes engine oil to drive a piston that is interconnected to the pitch change chain mechanism in the propeller hub. The cylinder the piston moves in is attached to the front face of the propeller hub. It was the cylinder and some of the hardware used to attach it to the face of the hub that were found in the separated spinner.

The airplane crashed in a near wings-level attitude on rocky terrain. Examination of the airframe and engine showed nothing significant to the cause of the accident. Records showed the propeller had accumulated 2,103 hours in service since being overhauled 15 years prior to the accident. The propeller manufacturer's recommended time before overhaul (TBO) is 1,200 hours. However, owners utilizing their airplanes for personal or private use are not required to adhere to the recommended TBO.

Examination showed one propeller blade was attached to the hub, but the other was missing. A diligent search for the missing blade was fruitless. We did not believe the missing blade had separated in flight because the engine was still on the airframe. Experience has proven that the vibrations accompanying a propeller blade separation while an engine is developing significant power will usually tear an engine right off its mount structure. However, as with any investigation, we always tried to limit the possibility of later complications by locating all the major components.

We realized the vibrations accompanying a propeller blade separation would have immediately grabbed both occupant's undivided attention. Accordingly, if the survivor noticed no unusual vibrations before the crash, it would be reasonable to assume the blade had not separated in flight. The FAA and a representative from the airplane manufacturer were participating in this investigation. In an effort to resolve our concerns about the missing blade, we visited the hospitalized survivor, hoping he would be able to answer several questions.

We weren't permitted to interview the survivor personally because of the seriousness of his injuries. However, through discussions with a family member, we learned the survivor was conscious and orientated as to his surroundings. We provided a list of several questions that could be answered with a "yes" or "no", and the relative came back with the following synopsis of the survivor's recollections:

> **The survivor thought they had blown an oil line because oil covered the windshield. He reported the pilot was struggling to regain control all the way down and the survivor did not recall any unusual vibrations or shaking of the airplane before the impact.**

As it turned out, the missing blade was found on a subsequent visit to the crash site. Propeller components were forwarded to the Safety Board's laboratory for metallurgical examination. The metallurgist's report showed six of the screws that attached the cylinder to the hub contained fractures indicative of fatigue. All other fractures on the propeller hub and pitch change components were due to impact forces sustained during the crash sequence and were not related to the cause of the accident.

Just what is fatigue? - When a metal part is placed under a steady, static load of less than the known limit strength of the metal, the part should theoretically last forever. If, however, the part is subjected to repeated or fluctuating loads, it may fracture at a stress level far lower than that required to cause failure under static conditions. This phenomenon is known as fatigue and is the most common cause of primary failures of metal components. Laboratory tests have proven that a fatigue fracture is progressive in nature. After a number (often many millions) of cycles of stress, a small crack forms in the region of highest stress. Under continued stressing, the fatigue crack grows until the cross section of the part is reduced to a point where the remaining area will not sustain the operating loads. Thereafter, instantaneous separation of the part occurs.

Fatigue of the cylinder attach screws had occurred gradually over an extended period of time. The instantaneous final separation of all the screws still attached to the hub occurred because they had insufficient strength remaining to withstand the normal operating loads. Separation of the cylinder disabled the propeller governor, and allowed a spring installed in the hub to drive the propeller blades into full low pitch (the smallest blade angle producing minimum resistance), and this accounted for the rapid rise in noise intensity heard by the witnesses as the engine and propeller speeds increased. The observations of the survivor in conjunction with the other evidence obtained during the investigation provided the answers to the immediate cause of the crash. Based on the evidence, it was reasonable to arrive at the following conclusions:

a. The rattling the witness heard in the propeller spinner was caused by screws used to attach the propeller cylinder to the hub that had separated because of fatigue.

b. Oil leaking through the loosened cylinder gasket provided the source for the oil streak the witness observed on the left side of the engine cowling.

c. Inadequate maintenance and inspection of the propeller were involved in the accident.

d. Engine oil utilized for propeller control dispersed onto the nose and windshield, causing some visibility restrictions. However, loss of the propeller cylinder did not cause any failures or malfunctions in the flight control system. Accordingly, it was considered the loss of control occurring after separation of the propeller components resulted from spontaneous reflex actions by the pilot that were without logical explanation.

The Safety Board requires that its investigators return aircraft records and any components retained for testing, laboratory analysis, etc., to the registered owner of the airplane. Following telephone conversations with the pilot's spouse, I elected to personally deliver the records and propeller components we had retained. This was a practice followed on those occasions where the travel distance wasn't too great and it was sensed that a personal visit might be warranted. She was a very gracious lady, residing in their large, custom-built home adjacent to the runway. Following our in-depth discussion of the accident, she showed me photographs of her husband in the company of Hollywood personalities and other prominent citizens.

She told me how he had been enjoying an exciting new career connected with golf, his favorite pastime, since retiring from his airline flying assignments. She showed me through the large hangar and workshop between their house and the runway. The hangar accommodated both the airplane involved in the accident and a small sport airplane the pilot owned. The workshop area was used for general maintenance and his venture into the golfing arena. Based on the forthright attitude she exhibited, it was obvious she was making progress with the adjustments necessitated by the circumstances of the accident. I remember her commenting how we all must be prepared to go eventually; however, she felt her husband was cheated out of 10 to 15 years of happiness in his retirement.

Reflecting back to the story of the religious man on the rooftop that precedes our discussions of this accident: St. Peter might have asked the pilot why he had ignored the warnings given by way of the rattling noise in the propeller spinner, and the oil streak down the side of his airplane.

- -

EXERCISED POOR JUDGMENT

The title above is about the worst thing that can be said about a pilot's headwork when using the Safety Board's manual to code the causes of accidents for computer input. However, it is considered some accidents are caused by such atrocious headwork by pilots that it goes far beyond the mere exercise of poor judgement. Call it what you may, but I believe foolish, irrational, moronic, or stupid, are more appropriate adjectives in some instances. However, when my below-listed analysis of the evidence gathered during the course of an investigation arrived at the Safety Board's headquarters in Washington, it created quite a ruckus.

The pilot made inadequate preflight preparations when he attempted the flight in the airplane with hardly

> **enough fuel to make a circuit around the airport. Poor judgment is evident; however, with his prior knowledge of the quantity of fuel on board at the time of the takeoff, his actions bordered more on stupidity."**

The tone of voice used by the chief of our analysis branch when he called, showed the wording of my analysis had not tickled his funny bone. However, I considered the wording used did not require an apology. I mentioned it was part of a restricted memorandum that would not be available to the public. My suggestion was that they either ignore my comments, or send it back for a rewrite. The caller went on to inform me the use of such terminology was totally inappropriate in any Safety Board correspondence. We terminated our discussions after I promised never ever to commit such a dastardly deed again.

Listed below are synoptic resumes of several accidents where it was considered the pilots exercised extremely poor judgment. Even though now retired, I am living up to my vow to never call a spade a bad name. Accordingly, readers may determine for themselves the pilot that wins the booby prize for exercising the **"Poorest"** judgment.

- -

BUT I CALCULATED I HAD A FUEL RESERVE OF NINE GALLONS

The total fuel quantity available in a relatively high performance single-engine airplane was sufficient for a flight of about 4 hours 30 minutes' duration at the optimum cruising speed. The pilot was conducting a round-robin flight to a destination more than two hours away. He reported the airplane was refueled to capacity before the initial takeoff but admitted he did not follow the good operating practice of visually checking his fuel supply before either takeoff.

The outbound leg of the flight was conducted without incident. However, on the return leg, the engine suddenly stopped about 30 miles short of the destination. It was shortly after midnight, but the pilot managed a successful wheels-up landing on a highway. The pilot and passenger were not injured, but the airplane received even more damage when it was struck by an 18-wheel tractor-trailer that came rolling down the highway.

Investigation showed the engine failure was caused by fuel exhaustion. This was an occurrence where we had a pilot who had not visually checked his fuel supply, sitting there looking at fuel gauges reading empty, but still relying on that nine gallons of fuel reserve he had calculated for the round-trip flight. The occupants were on a business trip that provided ample opportunity for the airplane to be refueled before the return flight. This occurrence is an example of another avoidable accident.

Accidents resulting from total exhaustion of the fuel supply have plagued the aviation community since the days of the Wright brothers. There is just no rationale for such occurrences, and the pilots are left defenseless. The airline distance from Jacksonville, Florida, to Miami, Florida, is about 300, miles and the aeronautical chart shows more than 24 airports along the route that are suitable for light airplanes. That averages out to an airport about every 15 miles. Pilots operating under visual conditions can usually make a refueling stop in less than 30 minutes. However, it appears there is nothing that can be done to teach pilots that aviation fuel will not stretch.

JUST FOR THE FUN OF IT

Flying across the countryside at high speed just above obstructions is fun. The terms "flat-hatting" or "buzzing" are sometimes used to describe such activities. During my Marine Corps career, I did a lot of it, some authorized, and some not. The day in 1951 is well remembered, when I telephoned my mother who was living in the family farmhouse her Daddy had built outside Farmville, Virginia. She was asked to be on the lookout because we were going to pay her a visit in our jet fighters in about an hour. At the time, I was a master sergeant flying in a two-plane section with Major Nate Peevey, a fellow aviator who could fly rings around me. We made the initial low pass to get Mom's attention. After she came out in the front yard, we made six more passes in a tail chase formation with me in the lead fighter. After our final low pass, we rocked our wings to wave good-bye before heading back to our base in North Carolina.

Upon returning to my residence that evening, I called my mother to see how she liked our show. She said she wasn't startled because she was expecting us and was thrilled by our exhibition. However, she reported our demonstration had received mixed reviews from her neighbors. Some realized exactly what was transpiring and enjoyed it as much as she did. But her description of the manner in which our antics had affected other neighbors was so shocking to hear that I was wary we might be in for an official reprimand for disturbing the peace.

Back in 1951, we didn't have all that many jet fighter planes. For those who have never experienced it, a jet airplane making an unexpected low pass at high speed directly overhead is a frightening event that will cause anyone's heart to jump in their throat. That's because we don't hear them until they are directly overhead; then, they're gone in a flash, and the sudden roar can be very startling. Mom related the following:

The mule a worker was using to plow tobacco on an adjacent farm bolted and ran away. The field hand, who was almost scared to death, knelt in the tobacco rows and prayed because he thought the judgment day had surely come. Other neighbors were almost frightened out of their wits because they couldn't possibly imagine what was happening that could create so much commotion in the relatively tranquil, rural countryside. All manner of confusion was created in the barnyards with chickens, ducks, and geese flying about, hogs roused out of their slumber, and other farm animals acting up and cavorting about. To make matters worse from my standpoint, the word spread like wildfire that my Mom's son, who was in the Marine Corps, was one of the culprits. But we lucked out, because we never heard anything through official channels about our escapades.

I hardly ever visit relatives in Farmville, Virginia without being reminded of this episode. At a family reunion in 1989, some 38 years later, several relatives reminded me of the event. This turned out to be the apex of my flat-hatting career. Realizing nothing could ever outdo it for sheer excitement, I cooled down my buzzing activities.

While researching my files, it was surprising to learn that slightly more than five percent of all my investigations involved pilots who misjudged their clearance while skimming across the landscape, just for the fun of it. In view of the relatively high speeds involved, the results were always tragic. In many instances, the gravity of the situation was magnified because the crashes occurred while family members or close friends were watching. One occurrence involved a pilot flying so low he killed his brother. The victim was watching the approaching airplane with a friend while standing in a pasture. Sensing that the plane was too low, the victim's friend hit the deck; however, the pilot's brother remained standing and was struck by the propeller. There was no damage to the airplane and the pilot landed safely.

UNAUTHORIZED LOW FLYING

After a concert in Knoxville, Tennessee, the rock band boarded a customized, Greyhound-type bus for a trip to Orlando, Florida. After an all-night drive, they stopped at a residence alongside a private airport in central Florida. The bus driver used a single-engine airplane to give local flights to some of the musicians and others associated with the band. He took off with two band members on board and made several low passes over the bus before landing and enplaning two other passengers.

After taking off on the second flight, the pilot made several more low passes. The bus was parked in front of a spacious, north-facing, Georgian style residence. On the final low pass, the airplane was in a steep left bank while turning towards the side of the bus. The pilot misjudged his clearance, and the left wing separated when it collided with the side of the bus. The remainder of the airplane crossed over the top of the bus and severed a large pine tree before crashing through the roof of a two-car garage attached to the house. The airplane was destroyed and the three occupants were killed. Postcrash fire erupted that destroyed the garage and two automobiles parked inside. The flames spread rapidly, causing significant damage to the living areas in the residence.

The owner of the airplane stated it was being stored in a hangar at the airport and no one had permission to fly it. Such occurrences pose massive problems for those suffering losses or damage. That's because there may be no one who is culpably responsible, from whom damages might be collected, unless the pilot possesses substantial assets. However, claimants will experience very little difficulty locating liability lawyers who are willing to go looking for a deep pocket, the term they use to imply a money source, provided the lawyers receive a large percentage of any judgments that may be rendered. In my hometown of Melbourne, Florida, they were advertising their services on TV daily with the statement, "No recovery, no costs to the claimant" until the Florida Bar initiated restrictions against such practices.

As exemplified by crop dusting flights, some low flying is necessary. When accidents occur while pilots are flat-hatting and not endangering the life or property of another person, it is referred to as unwarranted low flying. Pilots engaging in low flights that endanger the life or property of another person are conducting unauthorized low flight in violation of the FAA regulations. However, that provides little solace to those who are grieved, and others who have suffered losses because of such violations.

- -

IT'S OKAY, WE'LL JUST SNEAK IN UNDER THE CLOUDS

However, on many occasions, such flights may end up sneaking into radio or TV antennas, power lines, rising terrain, or some other obstruction. The accidents occur because some pilots are willing to risk their, and their passenger's, lives by attempting visual flight while the ceiling and visibility are well below the prescribed minimums. The fact that pilots attempting such flights are in violation of FAA regulations doesn't prohibit them from trying. As exemplified by occurrences where airplanes collide with power lines, there are occasions when pilots are forced to fly extremely low in order to maintain visual contact with the terrain. Pilots attempting such flights sometimes attempt to navigate by what is called the asphalt compass, meaning just above a highway; or the steel compass, meaning just above a railroad.

Retirement from the Marine Corps brought a seven year halt to my flying activities. Upon commencing to fly light airplanes again, a decision was made to conduct the flights under visual rules because I didn't think enough flying would be accomplished to maintain instrument flight proficiency. It took only one flight from Florida to Tennessee in marginal weather conditions to show the fallacies in my decision. Flying around under the clouds when the weather was marginal proved to be a new and harrowing experience. Upon returning to Miami, Florida, flights were scheduled with an instructor pilot and I was instrument-qualified before my next cross-country trip.

Three fatal occurrences involved pilots who flew their airplanes into guy wires supporting radio or TV antennas. Some of those antennas are more than 600 feet tall. In all three instances, witnesses reported the tops of the antennas were not visible because of the fog. In one case, sheriff's deputies arriving at the scene several minutes after a collision reported only the lower one-third of the antenna was visible. All of the antennas were depicted on Sectional Aeronautical Charts which light airplane pilots conducting visual flights should either possess, or study before taking off.

Twelve of my investigations were fatal occurrences in which the pilots collided with power lines they didn't see because the visibility was restricted. In three of the cases, the pilots flew their airplanes straight into the wires while in level cruise flight. In the other accidents, they descended into the power lines while attempting to land at uncontrolled (no control tower) airports.

My assignments included a dozen investigations where pilots flew their airplanes straight into rising terrain. Most of these accidents occurred in Georgia, Tennessee, and the Carolinas. Florida, with its pancake-like topography, was not involved. The circumstances of such occurrences are relatively easy to explain when you can stand at the crash site and observe that the airplane severed trees at progressively lower heights on the up-sloping terrain before crashing into the side of a hill or mountain.

The most difficult phase of some of these investigations involves just getting to the scene. One rainy day, U.S. park service personnel accompanied an FAA inspector and myself on a rather extended trek up the Appalachian Trail. About two hours after arriving at the crash site, located several hundred feet below the crest of a steep mountain, the park personnel informed us we would have to spend the night in the mountains if we didn't start back within the next 30 minutes. We replied we were almost finished. Upon starting back, they asked what we had learned. I replied: "Well, we already knew the pilot was on a visual flight and the mountaintops were enveloped in clouds. It's clearly evident he didn't see the mountain because his path through the trees shows he was in level flight at the time of the crash. Beyond that, I reckon the most significant thing that's been learned, is the impact forces didn't budge this mountain ridge one iota."

FLAT OUT DANGEROUS

The FAA requires that all pilots take a flight check with a certificated flight instructor every two years. Officially, it is called a biennial flight review. Upon satisfactory completion, the instructor makes an endorsement in the pilot's logbook. About a year before I retired from the Safety Board, the biennial review was scheduled with a young flight instructor in a light, twin-engine airplane. Before starting the engines, I requested that we not simulate an engine-out emergency while doing slow flight maneuvers. He asked why, and I responded we didn't want to get into an inadvertent flat spin from which we might not be able to recover. The instructor went along with my request, and endorsed my logbook at the conclusion of our post-flight debriefing.

He expressed some puzzlement as to the reasons for my request regarding simulated engine-out emergencies. Thereafter, we discussed the accidents I had investigated where the evidence showed the airplanes were in flat spins when they impacted the ground. We covered the possiblity of entering an inadvertent flat spin while in the slow flight regime, and the fact that it might sometimes be impossible to effect a recovery. It was learned the instructor had about 400 hours' pilot time. At the conclusion of our chat, he readily admitted he was not fully aware of all the inherent dangers.

There are no restrictions against spinning many light, single-engine airplanes. However, spins are not authorized in most other models because of structural limitations. In a normal spin, sometimes called a tail spin, the pilot has some airspeed. This provides airflow across the airfoils and flight control surfaces as the airplane descends in a nose-down attitude while turning about its longitudinal axis. In order to effect a recovery, the pilot uses the aileron and rudder controls to stop the spinning motion; and then employs the elevator control to recover from the dive after attaining an appropriate airspeed.

When in a flat spin, the airplane is turning and descending with the wings and fuselage in a near level attitude. In the event the pilot has some airspeed, he may be able to effect a recovery through use of the flight controls and engine power. However, there are occasions where pilots have little or no airspeed after finding themselves in a flat spin. In such instances, it may be impossible to effect a recovery regardless of the altitude when the airplane starts spinning. Under such circumstances, impact with the terrain is unavoidable.

Standing procedures in the Safety Board's Regional Offices call for the standby investigator to proceed immediately to the scene upon notification of a fatal accident. In those instances where investigators found the wreckage was not immediately recoverable, they normally returned to the office until salvage arrangements were made. Such was the case shortly before my retirement when one of our newly hired investigators, with very creditable qualifications, was preparing to return to the scene of an accident involving two fatalities.

Noting his preparations to leave, I asked what had happened. He replied arrangements had been completed to recover a light, twin-engine airplane that crashed in the Gulf off Florida's west coast. When I asked about the weather, he said it was CAVU, the acronym for ceiling and visibility unlimited. He related it was a dual instruction flight with two commercially-rated pilots on board, one undergoing training to upgrade to an airline transport pilot certificate. My response was: "Jeff, you've probably got a flat spin occurrence. If you are able to recover the wreckage, take a picture of the bottom of the fuselage before the crane lowers it onto the barge. You will be able to see every rib and stringer in the lower framework because, when in a flat spin, the airplane just pancakes into the water with very little, if any, forward momentum. If the victims are recovered, the autopsy protocol will show compression fractures of the vertebrae."

Upon returning to the office several days later, Jeff displayed the pictures of the bottom of the fuselage. There was no longer any doubt the airplane had impacted the water in a flat spin. The telltale deformations were clearly evident. Our analyses were further substantiated when both autopsy protocols showed compression fractures of the vertebrae.

Readers might wonder how the circumstances could be so accurately described from such sketchy information. At the time, I had investigated more than 550 civil accidents for the Safety Board, including several flat spin occurrences. There are few fields of endeavor where the saying, "There is no substitute for experience," is more appropriate than to that of the aircraft accident investigator. For flat spin occurrences, it might even be said that when you have seen one, you have seen them all. That is true, because the damage to the airplanes always follows such distinct patterns. Witnesses invariably report the engines were cutting in and out. That is something to be looked into; however, the experienced investigator knows the engine noises heard by witnesses are greatly influenced by their position relative to the engine exhaust stacks. As the airplane spins, that relative position is constantly changing. The engines sound loudest when the exhaust stacks are pointed at the witness and are less noticeable 180 degrees later when the nose of the airplane is pointed at the witness. Engine failures or malfunctions were not involved in any of the flat spin occurrences I investigated.

It should be understood that a flat spin is not something likely to occur while flying from Timbuktu to

Podunk Junction or between other locations. That's because pilots do not routinely fly or maneuver their airplanes at the slow airspeeds necessary for the airplane to enter a flat spin. Secondly, when the airplane gets in a flat spin, it is invariably inadvertent. All of my flat spin investigations occurred on dual instruction flights. Although there were no survivors, it is reasonable to assume the occupants were practicing low-speed maneuvers, because that is the time the airplane has the propensity to enter a flat spin.

Once the airplane enters a flat spin, it can be extremely difficult, if not impossible, for the pilot to effect a recovery. After entering the spin, the forward speed can decrease rapidly regardless of the pilot's response. This results in the airplane descending almost vertically. Under such circumstances, there is little or no horizontal air flow over the flight control surfaces, and movements of the cockpit flight controls are totally ineffective.

The number of flat spin occurrences we continue to experience shows there are other multi-engine flight instructors out there who are not fully aware of the inherent dangers associated with the phenomenon. Lacking complete knowledge, they give their students simulated emergency situations while in the low-speed regime. In some instances, this will actually cause the airplanes to enter flat spins.

One of my investigations involved a multi-engine flight instructor who had recently won the FAA's coveted "Flight Instructor of the Year Award." As part of the investigation, other students were interviewed who said the instructor would have them get the airplane all dirtied up, meaning with the landing gear and flaps extended, then while maneuvering at slow speed, the instructor would cut power on an engine to simulate an engine-out emergency. Without going deeper into the subject, that is the worst possible scenario insofar as an inadvertent flat spin is concerned. Instructor pilots employing such practices are jeopardizing their own and their student's lives.

Well written information and warnings have been provided in FAA publications, Safety Board news releases, and magazine articles. However, accident statistics reveal we still have pilots who are unaware of the inherent dangers. Studies by the FAA, pursuant to Safety Board recommendations, have shown there are no inadequacies or deficiencies in the regulations, and the airplanes fully comply with prescribed airworthiness standards. It must be remembered, there are no requirements for the FAA to revise existing airworthiness standards for the manufacture of any airplane model solely because of occurrences resulting from instances of extremely poor airmanship.

The FAA has rightly taken the position that the practice of demonstrating stalls with high power on one engine, and low or idle power on the other in the light, twin-engine airplane training environment is potentially a high-risk maneuver. The FAA has also established a policy whereby single-engine stalls are not a requirement for satisfactory completion of multi-engine flight checks. Accordingly, it should be apparent our flat spin problems are related to the actions of over-zealous flight instructors who attempt to provide instruction outside the required parameters.

Upon reading this chapter, one of my editors with no aviation experience wanted to know why something hasn't been done to put a stop to flat spin accidents if the factors involved are so easily recognized. Although unintended, he had zeroed in on the dilemma facing regulatory bodies and others striving to improve safety in any mode of transportation, or any other field of endeavor for that matter. Alluding to human nature, it sometimes seems every generation wants to be allowed to make its own mistakes, while shunning lessons that might be learned from similar mishaps in the past. And everyone will probably agree, when you try to legislate or initiate practices that go against basic human behavioral patterns, you have truly grabbed a tiger by the tail.

Some single-engine airplanes will also flat spin. The spins generally occur when the airplanes are allowed to stall after pilots load them in a manner providing a tail-heavy condition. Additionally, while probably not widely known or publicized, except in the popular movie, "Top Gun," some of our latest military jet fighters will enter flat spins from which recovery is impossible. The pilots generally escape through use of the ejection seats; however, at 30 million or more of our tax dollars per fighter, that gets mighty expensive.

It might be beneficial to review the overall experience level of some of today's flight instructors. This is not an attack upon, or an attempt to defame the ability of, any group or individual. However, it is important that everyone understands how the system works. When I started flying back in the early 1940's, my civilian flight instructors had grey hair around the temples and many years of flight experince to draw upon. Times were tough, and in many instances they owned the operation, managed the airport, pumped gas, performed aircraft maintenance, and flew charter flights. Then in an effort to keep the wolf away from their doors, they used their spare time to teach aviation ground school subjects or give flight instruction. Without question, the aviators of that era had a wide reservoir of aviation knowledge and experience to draw upon, and were eminently qualified to give flight instruction.

We still have many excellent career flight instructors out there who are happy with what they are doing. But, as we look back some 50 years later, it is apparent that some significant changes have taken place. Like everything else, flying costs have skyrocketed. In today's market, most job applicants are required to have in excess of 500 flight hours before they can expect to find employment as pilots. However, with proper training, they can acquire a flight instructor's certificate after amassing only 250 total flight hours. And that's what they do in many cases. No claim is being made that any of these low-time flight instructors are unqualified. The practice whereby recent graduates of flight schools initially become flight instructors is cited to show they do not have a great deal of flight experience to draw upon. In reality, most of them have visions of becoming highly paid airline captains. As for their flight instructor assignment, that's just their next step up the ladder as they strive to attain their goal.

Pilots undergoing training can learn much in the classroom. Likewise, they can become familiar with equipment and operating procedures in flight simulators. These are important training devices; however, meaningful practice in aircraft is the best way to build flight experience. Our military services do a far better job than civilian flight schools when it comes to identifying, and washing out of flight programs those students in the below average category, plus those not imbued with an appropriate degree of aeronautical aptitude. However, some do slip through the net.

Pilots are like any other representative group of people engaged in a common endeavor. On average, there will always be an appropriate percentage of Thinkers, Drinkers, and Stinkers. About the top ten percent will be outstanding; the next 60 to 80 percent will be in a broad average group with an ill-defined base; and the remainder will be below average. The outstanding individuals, in company with a small percentage from the average group, will perform admirably from day one. They were almost born with their hands and feet on the flight controls, quickly assimilate aviation knowledge, and make rapid progress. Pappy Boyington of the Marine Corps and Chuck Yeager of the Air Force are notable examples. There are many others; it would require several pages to name the Marines I knew who were in this group. Categorizing myself, I would probably fall into the bottom third of the average group. Fortunately, I recognized my limitations and kept on learning; however, I was never a candidate for that top ten percent.

"Crew matching" is a program that has long been in use. Basically, it relates to the practice of always scheduling well qualified copilots to fly with "weak" captains. Some air line companies, and possibly some unions representing airline pilots, might claim such a procedure has never been used. However, reiterating what was said previously: "They're not all Chuck Yeagers and Pappy Boyingtons, they are just a group with the usual percentage of Thinkers, Drinkers and Stinkers."

"Perfection is obtained by slow degrees; she requires the hand of time." Voltaire, French writer and philosopher.

THE GREMLINS DID IT

Most of the mishaps we have been discussing were caused by pilots. Let us now direct our attention to some accidents where pilot factors were not involved. Among that group we will find some real crowd killers. Many were caused by improper maintenance and inspection procedures. Others happened because of inadequacies in the design process. Some individuals, including those ambulance-chasing liability lawyers, would substitute the term, "faulty design." However, "fault" denotes guilt to a degree involving culpability, and such a connotation is completely erroneous.

Voltaire's "hand of time" reference provides a ready explanation for the unforeseen difficulties experienced with airplanes after they have been placed in service. In many instances the FAA has to issue mandatory airworthiness directives to assure safety will not be jeopardized. Basically, they will require redesign, or more stringent inspection criteria, for various parts and components previously considered to be in compliance with stipulated standards.

Engineers have earned the right to be looked upon with compassion. Whether designing football helmets, baby buggies, or airplanes, they are are hounded forever by that never-erring second-guesser, the test of time. Initially, they must compute the normal loads and stresses a particular part or component is expected to carry. Next, they incorporate all the variables that might affect the normal loadings. Thereafter, they contemplate and make allowances for the abnormal, unconventional, out of step, and just plain abusive stresses various users might exert. Surely, there are other considerations, so, at the conclusion of their computations, they throw in about a 50 percent safety factor with the hope it will account for unforeseen or unexpected circumstances. Admittedly, this is a rather nonscientific discussion of the design engineering process; however, it is probably not too far off the mark.

Take the Boeing 737 airplane that became a convertible model when an 18-foot section of the cabin top blew off while on an intra-island flight in Hawaii. Seven passengers and one flight attendant were seriously injured. The rapid decompression of the cabin swept another flight attendant overboard and she was never found.

The 19-year-old airplane had accumulated 35,496 hours during 89,680 flights. This correlates with an average flight duration of only 24 minutes. However, a far more important factor insofar as the accident was concerned was the fact that the pressurized portion of the cabin had been subjected to a pressurization and depressurization cycle on each flight.

The Hawaiian Islands environment is recognized as being susceptible to the effects of corrosion. Aloha Airlines took delivery of the airplane after it was manufactured in 1969. The top section of the fuselage separated after fatigue cracks formed in the areas where the various fuselage sections had been joined during manufacture. A representative fuselage section of the airplane had been tested to 150,000 pressurization cycles during certification. However, the "hand of time" showed the certification tests did not reflect the actual fatigue performance of the fleet aircraft. The Safety Board cited deficiencies in the company's maintenance and inspection program in its probable cause determination while listing other circumstances and events that had contributed to the accident.

The accident resulted in renewed study of the service life of today's high-performance jet airliners. The aviation community has come to realize that these airplanes are not going to give satisfactory service forever. Except for the Douglas DC-3 twin-engine transport that was manufactured in the 1930s, and will probably still be flying at the second coming of Christ, other transport models will have to be retired at some point in time. However, it will be possible to extend their service life through improved manufacturing procedures, more stringent maintenance and inspection criteria, and better corrosion control measures.

- -

"COLD TURKEY" ENGINE FAILURE

The pilot was the Commanding Officer of a squadron flying one of our most advanced jet fighters. A few months prior to the accident, he purchased an airplane registered with the FAA that he had been flying for personal and pleasure use. At the time of the accident, the pilot was ferrying his airplane to a location where an annual inspection was to be performed.

The substance of the pilot's account of the accident follows:

> **Visual weather conditions prevailed and the flight arrived over the uncontrolled (no control tower) destination airport without incident. The pilot overflew the field to check on the wind and other traffic before entering a right downwind leg for a landing to the north. He accomplished the routine prelanding checks and initiated the turn onto the base leg about a mile south of the airport. The engine stopped**

abruptly, without any coughing, spitting, or sputtering, as he initiated the turn onto the final approach.

The flight was about 400 feet above the terrain without a suitable emergency landing site. The pilot made an immediate decision to land in the treetops instead of attempting a 180 degree course reversal that might have permitted a ditching in a bayou. He reduced the airspeed while descending and stalled the airplane into the treetops. Thereafter, the pilot found "Good going until a big tree popped up."

The leading edge of the right wing collided with the tree about 35/40 feet above the ground. The airplane came to rest suspended in the tree with the fuselage partially wrapped around the trunk, the tail in a near upright and level attitude, and the left wing extending upward into the branches. The engine and right wing were dangling down, and the right side of the cabin was slit open. The seriously injured pilot was forced to remain in the cockpit until rescue equipment arrived. And all the while, fuel was spewing from the left wing tank into the cabin.

After checking and obtaining necessary documentation of the flight qualifications of the pilot and the maintenance history of the airplane, the thrust of the investigation was directed towards the factors relating to the engine failure. Authorization had been granted to salvage the wreckage, and it was moved to the airport before my arrival at the scene. After photographing the wreckage, documenting the cockpit instrument readings, and the settings of the various switches, valves, control levers, etc., we directed our attention to the powerplant.

The engine was cut away from the fuselage during salvage operations. After removing one of the magnetos from its mounting on the accessory section, it was observed that the drive gear did not rotate when the propeller was used to turn the engine crankshaft. By way of explanation, "the engine gear train was intact," is a phrase commonly used to describe the results of an engine examination. That means piston and valve action is found in each cylinder, and there is turning of the accessory gear drive train when the engine crankshaft is rotated. The engine examination in this instance showed the gear train was not intact.

Thereafter, further examination of the engine was suspended. Upon visiting the hospitalized pilot, he stated he had experienced absolutely no difficulty with the operation of the engine since purchasing the airplane. The pilot very much wanted to know why the engine had failed. He was advised arrangements had been made to ship the engine to the manufacturer's facility in Pennsylvania for a detailed teardown examination. I told him we were reluctant to further examine the engine locally for fear we might overlook, or destroy, significant evidence.

The engine was a 300-horsepower, state of the art, fuel injected model that had proven to be quite reliable. It had accumulated 2,093 hours in service, including 1,125 hours since overhaul by an Aircraft and Powerplants (A&P) mechanic; and 198 hours since replacement of the number two cylinder. Upon arrival at the manufacturer's facility, it was impounded until necessary arrangements were made to conduct our examination.

The investigative team was comprised of myself, one of the manufacturer's senior engineers, their metallurgist, and several engine mechanics. We found some loose metallic particles and evidence of excessive wear on many interior components. The engineer noted several discrepancies in procedures and practices used during overhaul of the engine and replacement of the number two cylinder. The sudden and complete failure of the engine occurred when the accessory gears stopped rotating. Close examination showed a foreign steel particle had become lodged between the gear teeth. This was sufficient to cause overload failure of the accessory drive gear mechanism.

The accident provided an example of the benefits derived when failed components are returned to the manufacturer's facility for examination. Normally, we accomplished about 90 percent of our engine examinations at the accident scene or nearby maintenance facilities. The decision whether an engine is to be shipped back to the manufacturer is based on a judgment call by the Investigator-in-Charge. In this instance, we had determined from our on-site examination that failure of the accessory gear drive mechanism had caused the engine failure. However, the photographic documentation of the steel particle lodged between the gear teeth would probably have gone undected if the teardown had been accomplished at the accident scene. Neither would we have uncovered all the nonstandard procedures employed during overhaul of the engine, and replacement of the number two cylinder.

There are several avenues open to aircraft owners when their engines are due for overhaul. It may be accomplished by the engine manufacturer; an FAA certificated overhaul facility; or a certificated powerplants mechanic. During my tenure with the Safety Board, I met many mechanics who did completely satisfactory engine overhauls. However, as symbolized by this mishap, engine overhauls by mechanics may be the avenue where the most potholes might be found.

- -

THE DAY EVERYTHING WENT WRONG

It was a nice clear day to go flying. The heavily loaded McDonnell-Douglas DC-10 passenger jet lifted off after a 6,000-foot ground roll down the 10,003 foot runway and climbed to a height of 325 feet while accelerating to an airspeed of 172 knots. The airplane then commenced decelerating and started a roll to the left after the left engine and engine pylon assembly separated from the wing. Despite the use of maximum flight control inputs by the flightcrew, the airplane continued to roll left until the right wing passed through the vertical plane. Thereafter, the nose dropped and the airplane crashed at a point 4,600 feet beyond the departure end of the runway. All 271 people on board the airplane, and two persons on the ground, were killed.

The accident involved American Airlines Flight 191 that was departing from Chicago, Illinois, for a flight to Los Angeles, California. Separation of the engine, in itself, did not cause the accident. The investigation showed the flightcrew flew the airplane in accordance with the prescribed engine failure procedures. When the engine separated, it severed some hydraulic lines. The severed hydraulic lines allowed the left outboard leading edge slats to retract. Other systems, including the leading edge slat disagreement (assymetry) light, and the stall warning system, were also disabled when the engine fell off.

The lift produced by the wings at a given airspeed is increased when the leading edge slats are extended. After the outboard leading edge slats retracted, the left wing stalled. The crew had no way of determining that the slats had retracted, or that the left wing had stalled, because the slat disagreement light and stall warning system were not operable after the engine fell off. Since the leading edge slats on the right wing remained extended, it continued to generate lift. The stalled condition of the left wing, in conjunction with the lift-producing condition of the right wing, caused the right wing to rise until the airplane rolled completely over to the left.

In summary, the loss of control was caused by the combination of three events: retraction of the left wing's outboard leading edge slats; the loss of the slat disagreement warning system; and the loss of the stall warning system. All were caused by the separation of the left engine and its pylon assembly. Each, by itself, should not have caused the flightcrew to lose control. However, happening together during a critical phase of flight, they created a situation which afforded the flightcrew an inadequate opportunity to recognize and prevent the ensuing stall of the airplane. According to the Safety Board, the root cause of the accident, "Separation of the left engine and pylon assembly," resulted from damage sustained because American Airlines personnel utilized nonstandard procedures when installing the engine on the wing.

- -

A MASSIVE BLOWOUT

Another tragic occurrence in which all 346 occupants perished involved a McDonnell-Douglas DC-10 airplane operated by Turkish Airlines. The flight had departed from Paris, France, bound for London, England. The airplane was filled to capacity because a flight by another airline had been cancelled. The accident occurred when the aft baggage door, that was not properly closed prior to the takeoff, separated on the initial climbout after the takeoff.

The baggage compartment is under the floor of the passenger cabin. Upon separation of the baggage door, there was an outrush of air as the cabin depressurized. However, because of design deficiencies, there were not enough openings in and around the cabin floor to allow the pressurized air in the passenger cabin to escape. The floor was never designed to be a pressure bulkhead. It was incapable of withstanding the pressure differential, and buckled downward. When the cabin floor gave way, it damaged the flight control system, causing the flightcrew to lose control. The decompression was so abrupt, six passengers sitting in two rows of seats located near the aft baggage compartment door were sucked right out of the airplane.

- -

ANOTHER BLOWOUT

United Airlines experienced an explosive decompression on one of its Boeing 747 airplanes when a forward baggage door separated in flight. The flight had departed from Honolulu, Hawaii, bound for Auckland, New Zealand, with 18 crewmembers and 337 passengers on board. Separation of the cargo door while climbing through 22,000 feet caused extensive damage to adjacent fuselage and cabin structure. Nine passengers were sucked right out of the airplane during the decompression and lost at sea. Engine numbers three and four on the right wing had to be shut down because of damage from foreign object ingestion.

The Safety Board initially found that the cargo door was not properly latched, and cited improper maintenance and inspection in its findings. After parts of the separated door were retrieved from the ocean floor, the Board revised its findings. It deleted the finding that the door was improperly latched because of damaged locks that should have been detected by inspectors. Instead, it found that the door latches were moved to the open position after takeoff because of a faulty switch or wiring in the door's electrical control system. The flightcrew managed a safe landing back at Honolulu. The "Fasten Seat

Belt" sign was illuminated at the time of the accident; however, access to the necessary rooms requires some movement about the cabin. At all other times, passengers should comply with the instructions by the flightcrew and keep their seatbelts securely fastened. Those intending to do otherwise would be wise to carry their own parachutes.

The "Tombstone Factors" discussed previously came into play for this mishap. Almost two years earlier, there was a cargo door opening incident on a Pan AM Boeing 747. Since that incident occurred after the catastrophic Turkish Airlines accident discussed above, the Safety Board considered that the FAA and Boeing should have reexamined the B-747 cargo door design. More positive door locking mechanisms were mandated for the DC-10 after the Turkish Airlines accident in France, and a more "fail-safe" cargo door locking mechanism had been required for the Lockheed L-1011 at initial certification. In reality, the earlier Pan Am cargo door incident turned out to be just another case of no tombstones, no timely corrective measures.

The Safety Board determined that the probable cause of the accident was the sudden opening of the cargo door and the subsequent explosive decompression. It cited the items below in its findings and conclusions:

a. A deficiency in the design of the cargo door locking mechanism.

b. A lack of timely corrective actions by Boeing and the FAA following the earlier cargo door opening incident on a Pan Am B-747.

- -

During my tenure with the Safety Board I investigated many other accidents caused by material failures on airplanes, from the tiny Piper Cub, to the giant Boeing 747. A small percentage were caused by design deficiencies. The majority were related to inadequate, or improper, maintenance and inspection procedures. On many occasions, the proficiency exhibited by the pilots in coping with the failures resulted in no injuries to persons on board, and no damage to their aircraft.

Almost every instance of material failure is traceable back to fallible mankind. The role played by us humans, who predominate in the design, manufacture, and operation of all our labor-saving devices, gadgets, machines, etc., explains why perfection will never be attainable. On the aviation scene and elsewhere we are always going to have to live with things that are going to break or otherwise fail to function. That being the case, prudence dictates that pilots have thorough knowledge of the emergency procedures for their airplanes; keep a cool head; maintain a safe airspeed (probably the most important consideration in coping with any emergency); and land as dictated by the circumstances. That might be immediately without regard to the terrain features below; at the nearest airport; or after continuing on to the original destination.

POTPOURRI

Efforts have been made to provide in-depth discussions of the wide range of circumstances and events that resulted in many aircraft accidents. The following medley relates to occurrences involving factors that didn't quite mesh with those reviewed in other chapters.

BEHIND THE AID BALL

It has nothing to do with Acquired Immune Deficiency. The ball we're referencing is an Aerial Ignition Device (AID). Those in the trade refer to them simply as an aerial ignition device or an AID; however, since the term "AID ball" is so descriptive, it has been adopted herein. Many people are aware that controlled burning is a measure sometimes used in today's land and forest management processes. The AID balls were developed to assist in lighting such fires.

The 1.25-inch diameter balls, similar in appearance to ping-pong balls, contain potassium permanganate. Ignition, with about a 30 second time delay, is achieved by injecting the balls with ethylene glycol or water-glycol solutions. Glycol is the primary ingredient used in automotive radiators for anti-freeze protection. During the on-scene investigation, the pilot's brother demonstrated how the ignition devices functioned by combining some fluid from the radiator of his pickup truck with the solution inside an AID ball. When the ball ignited about 30 seconds later, it burst into a white-hot fire that persisted for a few seconds.

The AID balls are shipped in lots of 1,000 each in a black plastic bag, inside a sturdy, square, one cubic foot corrugated box. The complete system consists of a supply of AID balls, a machine that primes the balls with the glycol solution before dispensing them overboard, a crewmember to operate the dispenser, and the pilot. The balls are systematically dispensed while the helicopter flies about three hundred feet above the terrain at a speed of approximately 50 m.p.h. The 30-second time delay assures that ignition occurs after the balls have landed.

The helicopter involved in the accident had a plastic bubble canopy. The door on the right side had been removed. The pilot flew from the left seat, and the dispenser was strapped to the floor in front of the right seat with the exit chute positioned outside the right door opening. The supply of AID balls had been removed from their corrugated cartons. The plastic bags of balls were loaded in the front portion of the canopy bubble forward of the instrument pedestal.

While the AID balls were being dispensed, an intense fire suddenly erupted in the bags of balls in the forward canopy area. The fire was so intense the crewmember climbed out onto the right landing skid. To further avoid the fire, he jumped, sustaining serious injuries, as the helicopter neared the ground. Because of the intense heat, the pilot lost control while attempting to land on a dirt road running through the forest, and the helicopter came crashing through the trees. The pilot sustained very serious impact and burn injuries before being rescued by forest management personnel who were nearby.

This turned out to be a very emotion-filled investigative experience. Any mention of this occurrence is sufficient to arouse the deep sense of compassion I felt for the pilot after the participating FAA inspector and myself donned the required protective garments and visited him at the hospital burn center.

In addition to dispensing the AID balls, the pilot owned and operated an aerial application business. He and his brother also grew some winter vegetables in a commercial venture. The accident occurred after one of those killing freezes that destroyed Florida's winter vegetable crop. It was apparent from discussions with the pilot and his brother that the freeze, in conjunction with the serious injuries to the pilot and the loss of the helicopter, constituted a serious financial blow. Although never in joint company with the pilot and his brother, I could readily sense the deep sense of brotherly love and concerns they shared for each other.

I harbored a very strong desire to accomplish anything that might be beneficial. And I never felt more frustrated in all my life. Because of the course of the investigation and the circumstances of the accident, it sometimes appeared I was working against the pilot. Before investigating this accident, I wasn't aware they used aerial ignition devices to start land and forest management fires. But as with many other topics, a fairly comprehensive understanding of the process was acquired before the investigation was completed.

The potassium permanganate-ethylene glycol incendiary device was developed in Australia. The use of the AID balls in North America came about through further refinement of the concept by Canada's Pacific Forest Research Centre in Victoria, B.C. The machine that injected the glycol into the balls before dispensing them overboard was manufactured by an engineering firm in Victoria, B.C. This firm provided much of the information required for the investigation during our telephone conversations, and forwarded technical manuals that proved to be very educational.

The pilot was required to obtain specific operating authority from the FAA before utilizing his helicopter to dispense the AID balls. The following was included in the limitations and restrictions stipulated by both the FAA and the manufacturer of the AID ball dispenser:

Securely fastened, extra cartons of AIDs may be carried in the cabin.

As stated earlier, the plastic bags of AID balls had been removed from the corrugated cartons before being loaded in the front portion of the canopy bubble. As in many other fire-related occurrences, impact and fire damage precluded a determination of the precise ignition source. However, in view of the intensity of the fire observed when a single AID ball was ignited, it was readily apparent why the fire spread so rapidly throughout the bags of balls.

There was no question the pilot violated his operating restrictions when he removed the bags of AID balls from their corrugated boxes. It was reasonable to assume the in-flight fire that precipitated the accident would not have occurred if the flightcrew had left the extra cartons securely fastened until it was necessary to open them. Both occupants stated they did not experience any difficulties with the operation of the helicopter. It was considered that an appropriate investigation had been accomplished, and sufficient evidence had been collected for the Safety Board to make its probable cause determination.

After my retirement from the Safety Board, a telephone call was received from a lawyer claiming he was representing the pilot. It was later learned he wanted me to revise my factual report to show the fire started when AID balls in the dispenser ignited, as opposed to erupting in the supply of AID balls stored in the forward canopy bubble. Such a revision would not be in agreement with the accounts of the accident we received from the occupants. Additionally, if that had been the case, the crewman should have jettisoned the AID ball dispenser. The installation instructions for the dispensor required the use of a quick-disconnect tiedown-strap so it could be quickly jettisoned in the event of an emergency.

The attorney wanted to take my deposition, but refused to pay my travel expenses. He had a copy of my investigator's factual report, and was informed my testimony would be limited to the factual evidence contained therein. In a final effort to be helpful, I reluctantly agreed to give a deposition over the phone.

The proceedings did not progress in an orderly manner. If still employed by the Safety Board, I would have called a halt to the deposition and telephoned our General Counsel for legal representation. I probably should have hung up on the lawyer, but thought I might be assisting the pilot. Standing firm on the facts in my report resulted in some of the most intense verbal exchanges experienced during my tenure with the Safety Board. Such proceedings are never expected to become confrontational, and the fact that they did in this instance still annoys me. I came away feeling my efforts to help the pilot had backfired miserably.

- -

AIRFRAME ICE

The effects of ice on an aircraft can be devastating. The ability of the wings to produce lift decreases, the gross weight and frictional drag increases, and ice clinging to the propeller blades causes a decrease in power. The results are an increase in the stall speed and a deterioration of aircraft performance. In extreme cases, two to three inches of ice can form on the leading edges of the wings in less than five minutes. Only one-half inch of ice is required to reduce by 50 percent the lift-producing capabilities of some wings.

Pilots may encounter two types of airframe ice. Rime ice, made up of opaque ice particles that take on the appearance of frost, will usually build up on the leading edges of the wings and tail components. Clear ice, having the appearance and consistency of the water frozen in our refrigerator icemakers, can build up on the leading edges of airfoils, and, in some instances, glaze over most of the exterior surface of an aircraft. Of the two types, clear ice is the most hazardous.

- -

AIRFRAME ICE AND WIND SHEAR, A DEADLY MIX

Accidents occurring because airplanes accumulate a load of ice are difficult to prove. In most instances, the ice has melted before the investigative team arrives. However, I was involved in the investigation of one occurrence where there was no question that airframe icing was involved.

The tragic mishap involved a family of five in which the parents and two sons perished, and the other son was injured. They had taken off from the airport serving their hometown for a weekend of skiing in Vermont with friends who were following in another airplane. The father was a well qualified, instrument-rated single and multi-engine, commercially rated, pilot. He was flying a high performance, six-place, single-engine airplane equipped with a 300-horsepower engine, and a 3-bladed, constant speed propeller. The airplane was not equipped with any anti-icing or deicing equipment. Like most airplanes, it did have a pitot heater installed to prevent ice blockage of the pitot-static tube used for airspeed indications.

The pilot had taken off about 10 minutes ahead of the airplane occupied by his friends. The pilot of the other airplane conducted his en route flight at a higher altitude, and landed at the destination without incident. The pilot reported: "I experienced negligible icing—none on the windshield and only a thin tread of rime ice on the leading edges of the wings." The pilot of the accident airplane received weather briefings on two occasions before taking off in the early afternoon. Upon checking the weather while en route, he was advised the report for Rutland was not available. When he discovered that Albany, New York was reporting an indefinite ceiling with 3/4ths of a mile visibility and light snow, he elected to land at Binghamton, New York, "and sit this one out."

But he didn't sit it out very long. About 20 minutes after landing, he telephoned the Elmira, New York flight service station and was again advised the weather for Rutland was not available. He was briefed on the area weather and given two pilot reports, one showing light to moderate icing, and the other showing light to moderate turbulence.

About 30 minutes later, the pilot called the Elmira flight service station again. He reported he had telephoned Rutland and they told him the airport had good visual weather with the ceiling above 3,000 feet. In part, the flight service station specialist stated: "He then told me he wanted to get any pilot reports I had on icing and tops, and any weather around the Rutland area, specifically Glen Falls, New York, and Burlington, Vermont. The manner in which the pilot told me about the Rutland weather, and the way he requested the tops and weather, displayed what I would term a belligerent attitude towards me and the previous briefer."

The specialist advised that the Rutland weather was still not available. Thereafter, the pilot was given a complete weather briefing, including flight precautions for both icing and turbulence from the area forecast. He was advised of a Burlington, Vermont pilot report for moderate rime ice; two Lebanon, New Hampshire pilot reports for light, clear, icing; and another for light to moderate turbulence. About an hour and ten minutes after he had landed at Binghamton, the pilot took off on an instrument flight plan to Rutland.

The flight contacted the Boston Air Route Traffic Control Center (ARTCC) and, while being vectored to the instrument approach fix serving Rutland, the pilot reported "we're getting moderate rime icing." The flight had been cleared to maintain 6,000 feet. After the pilot reported he was getting more icing and wanted to climb "to get out of this right now," he was cleared to climb to 8,000 feet. Thereafter, the pilot reported he was experiencing a problem; he was having a very difficult time climbing, was maintaining 6,000 feet, and wanted a course to clearer weather, and if necessary, back to Binghamton.

The situation was somewhat complicated because Boston Center had lost radar contact with the flight. The center controller requested radar assistance from Burlington approach control, and they established radar contact with the flight seven minutes later. After being advised Burlington had 3,500 feet overcast skies, 15 miles visibility, and radar contact with the flight, the pilot requested vectors to Burlington. The pilot reported he was continuing to accumulate ice, was unable to climb, was trying to maintain 5,000 feet, but radioed: "I just can't do it." He then declared an emergency.

A tense period followed because the flight was still over mountainous terrain and the lowest altitude the controller could assign was 5,500 feet. The controller advised the pilot the weather was good at Burlington and as soon as they got him over the mountains, they would descend him below the clouds and the ice. The pilot replied: "I'm losing altitude very quickly though; do the best you can to help me out here." In an effort to get the flight away from the mountains more rapidly, the controller assigned vectors left of the direct course to Burlington.

The flight eventually got out of the mountains, and the pilot reported the ice on the windshield was definitely coming off after he descended below the clouds. The pilot still experienced some difficulty locating the runway because of ice remaining on his windshield. The tower controller increased the brightness of the runway lights to aid the pilot. After reporting the runway in sight, the pilot was cleared to land on runway 15.

The tower controller advised the surface wind was from 180 degrees at 22 knots while the pilot was on his final approach. Moments later, he observed the airplane execute a drastic dive and recover. The controller advised the wind was from 170 degrees at 21 knots. A few seconds later, he observed the airplane execute another dive to near treetop level before recovering. The tower controller advised the wind was from 170 degrees at 24 knots. The pilot replied: "I just experienced what I think to be windsh (incomplete word)." Thereafter, the controller observed the airplane enter another dive and disappear behind trees before he observed a fireball at the approach end of the runway.

The airport emergency vehicles had been alerted and were standing by near the approach end of the runway. A witness in one of the vehicles reported he observed the airplane execute three porpoise-like maneuvers. He said it abruptly descended about 150/200 feet before recovering about a third of the altitude lost. The maneuver was then repeated with about the same altitude recovery. It then commenced another rapid descent and the airplane impacted the terrain in a flat attitude 12 feet short of the overrun pavement. A ball of fire erupted that persisted a few seconds and the wreckage slid 300 feet into the overrun area before coming to a stop just off the left side of the runway pavement. The fire trucks that were standing by immediately foamed the entire area to prevent further fire.

It was considered the pilot lost control of the airplane because his approach speed was too low. This was an occasion where the pilot should have made a "hot" landing. Obviously, the airplane's stall speed had increased because of the airframe ice. Additionally, the pilot should have added a few extra knots because of the probability of an encounter with wind shear. The runway length of 7,807 feet was adequate to accommodate a landing well above the normal touchdown speed for the airplane. After the occupants had endured their ordeal in the ice-laden airplane over the mountainous terrain, it was inconceivable the flight could end so tragically. This was an accident that did not have to happen. However, similar occurrences can only be prevented by pilots, because aviation authorities are powerless to regulate against such accidents.

This occurrence provides an exellent example of the dangers associated with flying into areas where icing is forecast. Pilots of airplanes not equipped with anti-icing or deicing equipment should be extremely wary. Once ice accumulates on an airplane, it can be difficult to eliminate. This is especially true when the pilot does not have the option of either climbing or descending as experienced in this instance.

After my three tours of duty as an instrument flight instructor during my flying career, I wasn't too concerned when the preflight briefing showed instrument weather conditions prevailing over the planned route of flight, with the destination weather at, or near, the minimums. However, on those occasions when the briefer began discussing the probability of an encounter with icing conditions, I became a very attentive listener.

A review of the recorded communications with this flight demonstrates the methods air traffic controllers employ when called upon to aid pilots in distress. How could anyone possibly expect more out of the system than this pilot received? Based on my experiences, this was not an unusual happening. Instead, it just exemplifies the degree of cooperation and assistance our air traffic controllers routinely provide.

SELECTED UNSUITABLE TERRAIN

Taking off or landing on unsuitable terrain is a factor in many accidents every year. Both paved and turf runways can become unsuitable because of ice, snow, standing water, etc. Additionally, turf runways may become unsuitable if they are soft from heavy precipitation, or if not properly mowed. Often, pilots manage safe landings at off-airport sites, only to crash on an ensuing takeoff attempt because they failed to obtain flying speed, or other factors related to the length, or condition, of the takeoff path.

BUT I HAVE THREE OPERATIONS TO PERFORM IN THE MORNING

The pilot, his spouse, and their four children boarded the airplane. The first takeoff attempt into a five knot headwind was aborted. The pilot then taxied to the far end of the 2,750-foot strip before commencing a takeoff in the opposite direction. Witnesses observed large "rooster tails" of water rising aft of the airplane as it accelerated on the ground roll. The plane lifted off about two-thirds of the way down the strip, and turned slightly left towards pine trees standing off the side of the runway. The nose was observed to rise abruptly, as the pilot attempted to climb by increasing the pitch attitude. The airplane stalled, and fell back on the runway 65 feet short of the end. Continuing its path, it overran the runway before colliding with an embankment. The pilot sustained fatal injuries, his spouse and two of their children were seriously injured, and the other two children escaped with minor injuries.

The pilot, a medical doctor, was flying a sophisticated, six-place, single-engine airplane. Good weather prevailed, but heavy rains had been falling and there was standing water on the turf strip, where the grass was six to ten inches tall. A friend who inspected the condition of the runway with the pilot said he advised the doctor not to take off because of the standing water. His recommendation elicited the response by the surgeon concerning his scheduled operations.

The cause of the accident was clearly evident. The standing water and high grass on the strip significantly inhibited acceleration during the ground roll. The pilot pulled the airplane off before attaining a safe flying speed. It stalled when he pulled back further on the control column in an attempt to clear the pine trees that wouldn't have been in his takeoff path if he had remained on the runway heading. This was another tragic accident that could have been avoided.

There were no criteria the pilot could call upon to gauge the retarding effects of the high vegetation and standing water on his takeoff acceleration. Pilots are left with their own analyses in such instances, and an educated guess is about the best that can be hoped for. Additionally, for reasons that are not fully explanatory, there is a tendency by many pilots to pull their airplanes off the ground prematurely when taking off on short runways, or runways where other circumstances or conditions might indicate that a successful takeoff might be in jeopardy.

But pulling the airplane off prematurely is about the worst thing a pilot can do in such instances. That's because, once off the ground, the pilot must maintain an excessive nose-high (pitch) attitude just to stay airborne. Such a nose-high attitude inhibits acceleration, because more of the airframe surface is exposed to the relative wind, thereby increasing drag forces. As seen in another light, the airplane is not streamlined into the wind, and mushes along, gaining little or no altitude, because of the semistalled condition brought about by insufficient airspeed and the excessively high pitch attitude.

The pilot exercised poor judgement by yielding to his self-induced pressures to attempt the flight in order to meet his prior commitments. The circumstances were greatly aggravated when he boarded his family for the takeoff attempt. These assertions are not made in the realm of hindsight, second guessing, or Monday-morning quarterbacking. Instead, they are related to the fact that prudence demanded the passengers and baggage be transported by other means to a nearby airport with paved runways for pickup. Besides preserving the safety of the passengers, such actions would have decreased the gross weight of the airplane, which would have enhanced the probabilities of a successful takeoff.

There is another aspect of this occurrence that merits discussion. The six-place airplane was equipped with three rows of seats, with the middle two seats facing aft. Everyone would probably guess the two children who escaped with only minor injuries were occupying the aft facing seats. There is no refuting the fact that aft-facing passenger seats greatly inhibit injuries in survivable accidents. Yet, some 70 or more years after airline passenger service was inaugurated, we still have forward facing seats. Currently, parents are authorized to hold infants during takeoffs and landing on air carrier flights. The FAA has under discussion a rule-making proposal relative to the merits of requiring that the infants be securely strapped in their own seats for takeoffs and landings. However, if the FAA's proposal ever evolves into a requirement, both the infant and their parents will still be subjected to the whiplash type injuries caused by the rapid horizontal deceleration that occurs in many accidents.

The entire problem would be better solved by just turning the seats around. Nothing would be safer for infants than to be held tightly against the breast or chest of their parents seated in aft-facing seats. The primary reason opponents espouse for not turning the seats around is their claim people don't like to ride facing rearward. During my military career, I rode in lots of aft-facing seats. It never bothered me, nor was it apparent the rearward facing ride adversely affected the other passengers in any manner whatsoever.

"Hangar flying" is a term most folks are familiar with. It refers to a bunch of pilots lying and bragging about their flying exploits while telling other tall tales. Well, we accident investigators did essentially the same thing, except the term "hangar bashing" would be more appropriate since we would be discussing aviation accidents. These were "no holds barred" sessions. We habitually used the term "dodo bird" when referring to pilots who were guilty of senseless or totally irrational behavior. Obviously, this was opposed to the more stately definition by bureaucrats who would probably say something akin to: "Their actions being devoid of normal clarity."

We said a dodo bird was one that flew about with its head tucked back between its legs because it didn't give a whit about where it was going. Instead, it only wanted to see where it had been. Leaning on that definition, the time has come to start treating passengers as if they're a flock of dodo birds. By doing so, we will make giant strides in our safety endeavors, and the passengers will miss out on absolutely nothing. That's because at thirty thousand feet or so, there's not all that much difference in the landscape scenes of where we're going, and where we've been.

EPILOGUE

The Safety Board provides the following definition for an aircraft accident: "An occurrence associated with the operation of an aircraft which takes place between the time any person boards the aircraft with the intention of flight and all such persons have disembarked, and in which any person suffers death or serious injury, or in which the aircraft receives substantial damage."

Many accidents involve only injuries to persons, with no damage whatsoever to the aircraft. Encounters with clear-air turbulence are prime examples. On many occasions, occupants receive injuries as they are bounced off the ceiling or otherwise tossed about, while the aircraft receive no damage.

A tragic accident involved a couple returning to their home field late one evening. The pilot parked in front of their T-hangar with the engine idling while his spouse disembarked. She ducked under the strut of the high-winged airplane and was killed when she walked into the arc of the propeller.

Two of my investigations involved mishaps where ground personnel were killed when run over by the wheels of wide-bodied jet transports being pushed back from terminal gates. As evidenced by these occurrences, it is possible to have an "Aircraft Accident" when there are no injuries to persons on board and no damage to the aircraft.

We have completed our discussion of many of the factors involved in aircraft accidents. A few were omitted, and others will crop up in the future. Without giving the impression that the Safety Board or its investigators are "down" on pilots, it sometimes appears the latter are burning the midnight oil in their attempts to find new and exotic ways to bust up their flying machines.

Four appendices are included for those desiring more detailed information about the subject matter. Briefly, they are as follows:

APPENDIX "A" - AN INVESTIGATION WALK-THROUGH - Shows the step by step procedures employed for a rather involved investigation. It demonstrates the precautions used to ensure that the Safety Board maintains jurisdiction over the proceedings. This is extremely important when failed components are involved. The outcome of huge liability lawsuits can hang in the balance.

APPENDIX "B" - BIRTH OF THE SAFETY BOARD - Shows the evolution of the National Transportation Safety Board. A frank, no-holds-barred discussion of the personnel changes, policies, and procedures of the various presidentially appointed Chairmen. Reveals how the small, efficient organization is being dismantled by the bureaucracy.

APPENDIX "C" - THE AIR SAFETY INVESTIGATOR - Discusses the attributes of a qualified Investigator-in-Charge. Shows why fully qualified applicants are hard to come by. Substantiates the requirement for bringing on board candidates with extensive flight experience.

APPENDIX "D" - THE AUTHOR - An autobiographic sketch of my aeronautical background and investigative experience. The somewhat lengthy account is presented for the sole purpose of allowing readers to evaluate whether they consider I was qualified to write this treatise on aviation safety.

AN INVESTIGATION WALK-THROUGH

On average, a Safety Board investigator covers 25 to 30 aircraft accidents per year. About 80 percent are instant replays of prior occurrences requiring another deja vu investigation. However, investigators dive into the remainder with fervor, because they have now come upon something challenging; something they can sink their teeth into; something they might use as a vehicle to improve safety by recommending changes that will prevent similar accidents.

These are the occurrences that command the attention of the other investigators when they return to the office. It's not unusual for everyone to chip in as the factors are discussed in great detail. To determine the causes of accidents that might have been preventable through revised regulations, amended airworthiness standards, or other corrective measures, and then institute no remedial actions, would be foolhardy. Exhibiting such a do-nothing attitude would defeat the primary purpose of the Safety Board's investigations.

HEADS WE WIN, TAILS YOU LOSE

The above cliche relates to accidents in general. That's because, in the final analysis, it makes little difference whether the mishaps were caused by the pilots, erroneous acts by other individuals, material failures, or whatever. The results are the same: we end up with busted flying machines, broken lives, and shattered dreams.

The investigation of accidents caused by material failures is a fertile field for Safety Board accomplishments in the accident prevention arena. It is far easier to redesign a defective part or component than it is to reprogram all the nasty weather that keeps sweeping across the country, or determine why so many pilots are prone to take unnecessary risks. To gain insight into the techniques routinely employed, come along as we retrace the steps taken during the investigation of an accident caused by material failure.

NOWHERE TO LAND

This accident occurred after the pilot experienced a complete loss of engine power. It is selected because it provides an in-depth study into the Safety Board's modus operandi. We will also gain insight into the cares and concerns exhibited by company executives and other responsible personnel for the reasons why components they manufacture failed, while showing their eagerness to find appropriate solutions. All of which repudiates the claims of those liability lawyers that we routinely turn failed components over to the manufacturers for detailed study without ensuring the proper safeguards. The lawyers would have us conduct the necessary tests and research at private laboratories. But without question, the finest source of expertise is found at the factories where the failed components were manufactured. Another consideration relates to the costs of the tests, research, and experiments conducted as part of an investigation. The manufacturers give us access to both their facilities and their expertise without charge. In the event the Safety Board was required to pay for all such activity, it would necessitate a much larger appropriation from Congress.

In the subject accident, the new, two-place, single-engine airplane, used primarily as a trainer, was being ferried from the manufacturer's facility to a purchaser in Florida. Two well qualified, commercially rated pilots were on board, one with over 8,000 flight hours and the other with about 1,000 hours. It was a clear, dark night, and after a refueling stop at Raleigh, North Carolina, they took off at 8:05 p.m. bound for Savannah, Georgia.

The pilots did not file a flight plan, nor were they required to. At 9:17 p.m., the pilot radioed the Florence, South Carolina flight service station and reported they had experienced a complete engine failure. Equipment was not available at the Florence flight service station to record their communications with the flight. The statement by the specialist on duty included the following:

The pilot gave his altitude as 2,000 feet, said he was over an area where there were no lights, but reported several lighted towers ahead of his position. He stated he was southbound and had no chance of making the Florence airport. Upon being questioned as to the possible cause of the engine failure, the pilot stated he had experienced a dual magneto failure. He related he had fuel, good fuel pressure, and the fuel selector valve was on. At 9:20 p.m., the pilot reported his altitude was 1,600 feet.

The specialist had no further contact with the flight, and the whereabouts of the airplane and fate of the occupants were unknown. The loss of communications with the flight created a flurry of activity in the Florence flight service station. The following chronology shows some of the measures initiated:

9:21 p.m. - The South Carolina Highway Patrol was contacted and given the general location of the missing airplane.

9:35 p.m. - The Jacksonville, Florida Air Route Traffic Control Center was advised of a possible downed aircraft north of Florence and requested to have aircraft in the area check for an emergency locator transmitter (ELT) signal.

9:58 p.m. - The Rescue Coordination Center at Scott Air Force Base in Illinois, the organization responsible for the overall conduct of searches for missing aircraft over the continental United States, was notified. This was after the local Civil Air Patrol Unit had been alerted.

10:35 p.m. - After the pilot of an airplane reported receiving an ELT signal in the area of Darlington, South Carolina, the County Sheriff's office was notified, and they dispatched deputies to search the area.

10:40 p.m. - A teletype Alert Notice (ALNOT) advising the airplane was missing was transmitted.

11:20 p.m. - A searching Civil Air Patrol aircraft reported receiving an ELT Signal north of Florence.

8:20 a.m. Saturday morning - the pilot of an army helicopter sighted a downed aircraft nine miles north of Florence.

11:50 a.m. - The Army helicopter advised that a ground party had located the scene. It was determined the wreckage was that of the missing airplane and both occupants had been killed.

12:20 p.m. - The FAA's General Aviation District Office in Columbia, South Carolina was notified the wreckage had been located.

12:40 p.m. - The previous Alert Notice (ALNOT) message transmitted for the missing aircraft was cancelled. A Preliminary Aircraft Accident Notice was transmitted.

The above chronology represents the FAA's typical response in such instances. Their personnel deserve high praise for the devotion to duty exemplified by their actions. How could anyone expect more out of the system? They truly provide a great service to the aviation community, as do all other individuals and organizations engaged in search and rescue efforts.

Late Friday night, I was notified by the FAA'a Miami, Florida flight service station that the airplane was missing. In keeping with our procedures, no actions were taken pending the outcome of the search efforts. At 1:00 p.m. on Saturday, an inspector assigned to the FAA's General Aviation District Office in Columbia, South Carolina called to say the airplane had been found. He was told I would be on my way as soon as possible, and we agreed to use the Florence flight service station as a contact point. I departed from my residence in Miami at 4:30 p.m., took a commercial flight via Atlanta, Georgia, and arrived in Florence at 9:35 p.m.

I visited the flight service station and received a briefing on the circumstances of the accident that included the pilot's statement concerning a double magneto failure. Thereafter, I proceeded to the motel where the FAA inspector from Columbia was staying. Arrangements had been made for a sheriff's deputy to come by at 7:00 a.m. and accompany us to the wreckage site. We decided to delay our departure until an electrical continuity meter was obtained. I always attributed a great degree of significance to the last statements of accident victims, and was determined to check the magnetos before otherwise disturbing the wreckage. We learn best from our mistakes, and I was mindful of a recent investigation where we may have lost some significant evidence. Even though assisted by a highly creditable and experienced team in that investigation, we failed to document every step taken during an engine teardown examination, to our later consternation.

Much of my knowledge about mechanical matters was obtained through on-the-job experiences while employed by the Safety Board. We were fortunate in this instance, because the individual representing the FAA was one of the finest maintenance inspectors I ever worked with. The expertise and assistance he rendered during the on-scene phase of the investigation was invaluable.

The two magnetos installed on aircraft engines are small electric generators that provide the current for spark plug ignition. The magneto switch is equivalent to the ignition switch in our automobiles. The cylinders on aircraft engines are equipped with two spark plugs, each receiving current from a different magneto. As opposed to the single ignition systems in our automobiles, the dual ignition system in airplanes provides more efficient combustion of the fuel/air mixture while ensuring continued operation of the engines in the event one of the magnetos should fail. Aircraft magnetos are rotated by a gear arrangement off the accessory gear that is driven by the engine crankshaft.

The magnetos are isolated from, and not dependent in any manner whatsoever upon the aircraft's normal electrical system. The magneto switch has four positions; OFF, LEFT, RIGHT, and BOTH. As part of the pretakeoff checks, the pilot rotates the switch to assure that both magnetos are functioning.

The crash site was not easily accessible. The airplane went down in a remote, heavily wooded area adjacent to the Pee Dee River north of Florence. After colliding with the tops of tall trees, it nosed into the ground vertically, coming to rest standing on the nose and leading edges of the wings.

First, we checked the magneto switch for proper operation. The switch operated satisfactorily and we found continuity to both magnetos. Thereafter, we proceeded as in any other investigation: photographing the wreckage, documenting the cockpit instrument readings, switch positions, various selector valve settings, etc. The hourmeter showed the airplane had accumulated only 5.6 hours in service. After determining we could gather no other meaningful evidence until the aircraft was moved, we made arrangements to salvage the wreckage.

No obstructions were found in the fuel system. The fuel lines, fittings, valves, etc., were intact except for obvious damage sustained during the crash sequence. The engine-driven fuel pump was intact and operational. The electric fuel boost pump operated satisfactorily when connected to an automotive battery. The spark plugs showed a serviceable appearance, and the electrodes were not shorted or eroded. The engine gear train was intact. (Meaning we obtained proper valve action, piston movement, and accessory gear rotation when the propeller was used to hand turn the crankshaft.) Experience has proven an engine passing the above tests will operate if provided with a combustible fuel/air mixture and an ignition source for the spark plugs.

In view of the pilot's statement concerning the dual magneto failure, it might appear we were over-investigating the accident. Lessons learned from past experiences tell us to turn every stone, leave nothing to chance. While such measures make the on-scene phase of an investigation a more lengthy process, the extra effort can prevent many headaches and much agonizing in those instances where it is later determined the initial suppositions didn't pan out.

Both magnetos were removed from the engine accessory section. The left magneto was intact, but did not produce a spark when rotated. We were unable to test the right magneto because of impact damage.

After making arrangements to ship the engine to the manufacturer's facility in Pennsylvania for operational tests, I hand carried both magnetos back to our Miami office. The exhaustive on-scene investigation lasted three days and I arrived back in Miami Tuesday evening. My office supervisor had been briefed daily as to the status of the investigation. His response was, "Looks like you are onto something up there, Tom; don't come back until you have collected every thread of evidence that might be pertinent to the accident."

The on-scene phase of the investigation had been conducted by the FAA inspector and myself. The accident gathered interest when another of the same model airplane was found with two "dead" magnetos. Wednesday and Thursday were used primarily in telephone conferences with interested parties as arrangements were made to test the magnetos at the engine manufacturer's facility. I departed from Miami late Thursday afternoon and arrived at my Pennsylvania destination at 10:30 p.m.

Our investigative team had increased significantly when we met Friday morning. Present were three representatives for the engine manufacturer, five representatives from the airplane manufacturer, two representatives from the FAA, a representative from the magneto manufacturer, and myself. During our brief opening session, I cautioned them to go slowly, while documenting every action to ensure significant evidence was not overlooked or destroyed.

We had six magnetos to check, four that had been returned under the engine manufacturer's warranty program, plus the two removed from the wreckage. The five intact magnetos were installed on a test stand and found to be inoperative. Disassembly inspection showed all five magnetos had shorted capacitors. After the shorted capacitors were replaced with those removed from magnetos in stock, all five magnetos operated satisfactorily. The capacitor was also found to be shorted in the magneto removed from the wreckage, which could not be functionally tested because of damage sustained during the crash sequence. I departed shortly after noon on Friday and arrived back in Miami at 9:30 p.m., still hand carrying the two magnetos from the wreckage.

Initially, it was my intent to monitor the operational tests of the engine. However, with the realization that we had determined why the engine had failed, it was requested that the resident FAA representative at the engine manufacturer's facility monitor and report on the outcome of the operational tests after the engine arrived. He readily agreed, and later reported that nothing pertinent to the cause of the accident was found.

The early part of the following week was spent making arrangements for the investigative team to assemble at the capacitor manufacturer's facility. I departed from Miami Wednesday afternoon and arrived in Chicago, Illinois at 7:00 p.m. Present at our meeting at the capacitor manufacturer's facility on Chicago's northside Thursday morning were two representatives from the engine manufacturer, two representatives from the airplane manufacturer, four representatives from the magneto manufacturer, four representatives from the FAA, two representatives from the capacitor manufacturer including their principal executive officer, and myself. The magneto manufacturer's facility was located in a city west of Chicago. Their principal representative was accompanied by some of their chief engineers. They also brought a truck-mounted stand on which we could operate magnetos during our tests, experiments, and evaluations.

I still recall the opening remarks by the Chief Executive Officer for the capacitor manufacturer as he welcomed us to their facility. He stated they had been manufacturing capacitors and condensors since the early 1930's. They made tiny ones for use in electronic and avionic components, plus almost everything in between up to the huge ones used in hydroelectric plants. He was visibly disturbed that a failure of their product had resulted in a complete loss of engine power that ultimately claimed two lives. We were given use of their facility to examine and test, to destruction if necessary, as many capacitors as might be required. He encouraged us to seek advice from their engineers and technicians. We were told many of their operations were conducted in a controlled, dust-free environment, and compliance with their procedures for entering those areas was requested. He related they made every effort to manufacture components to the stipulated specifications, after which a randomly selected lot from each batch was extensively tested before the batch was released. We were asked to let him, or his engineers, know if we found any discrepancies whatsoever in the manufacturing process, or noted any procedures they might institute to improve product reliability.

Capacitors are an integral electrical component in magnetos. Very basically, the construction of those installed in the failed magnetos consisted of aluminum foil windings, separated by mylar dielectric windings, on a paper core. After the windings were made by hand, a terminal stud was inserted, the ends were soldered, the assembly was inserted in a metal housing, and the winding was encapsulated, with epoxy, in the housing. Rated at 200 volts, the finished product was about the size of the last two joints on our little fingers.

We observed closely as the technicians removed and examined the capacitors from the magnetos installed in the wreckage. We found no defects in the manufacturing process. Examination of the windings after removal from the housing showed very small holes in the mylar dielectric. The technician said the holes appeared to be "low voltage punctures" between the aluminum foil windings. The holes permitted a short circuit, which was sufficient to cause both magnetos to fail.

We spent all day Thursday and Friday trying to duplicate the failures, without success. We had to apply about 3,500 volts to cause a new capacitor to fail. The test results were not representative of the findings in the failed capacitors. Causing capacitors to fail by increasing the voltage burned huge holes through several layers of the windings. We destroyed a great many capacitors and kept that truck-mounted test stand running overtime while conducting every experiment any of the assembled experts could dream up. And all we ever came up with were blanks. We didn't arrive at a single finding we could relate back to the cause of the engine failure in the accident aircraft.

While in Chicago, we learned another accident had occurred in the same airplane model because of a dual magneto failure. I turned to the principal representative for the airplane manufacturer and said: "We've got to ground these airplanes." He replied: "I know that, Tom." I asked whether they wanted to initiate the necessary actions. He answered, "Why don't you do it."

I immediately called my chief in Miami and explained where we were and what we thought ought to be done. He directed me to call an official in our Washington, D. C. office, since I was the only Safety Board employee who knew all the facts, circumstances, and conditions of the accident. I called Washington and provided a comprehensive report of the status of the investigation. Within 24 hours, the FAA had suspended the airworthiness certificate for the airplane, which effectively grounded all airplanes of that model.

I departed from Chicago Friday evening, stopped off in Washington as they had requested, and arrived back in Miami at 8:00 p.m. Monday. My role in the investigation was essentially over. We had identified the problem areas, taken actions to prevent similar occurrences, and the

FAA was not going to reinstate the airworthiness certificate until necessary corrective measures were instituted. Later, the FAA issued an airworthiness directive, (compliance was mandatory) requiring that engines installed on the airplanes be equipped with magnetos having revised specifications.

Upon returning to the office, I received the expected ribbing from the other personnel concerning the manner in which I had "milked" the investigation to keep myself off the accident standby roster. A comprehensive "Investigator's Factual Report" still had to be written. However, all investigators manage their time schedules and sandwich such duties between subsequent investigative assignments.

At the conclusion of the Safety Board's investigation, the aircraft manufacturer used 30 airplanes from their flight line for exhaustive tests. They contracted with outside engineering firms for some of the tests. A big riddle was related to the realization that the magnetos were giving satisfactory service in many other airplane models. Another puzzling aspect involved the means used by the pilot to pinpoint the cause of the engine failure so accurately. Possibly, both magnetos did not fail simultaneously, and through use of the magneto switch he determined one was dead before the other failed.

This completes our walk-through of a typical investigation. The Safety Board utilizes the same procedures in its investigations of our catastrophic accidents that produce many tombstones, the difference being they divide the investigative activities into various working groups, each responsible for particular facets of the overall investigation. This "team" concept has been employed many years and has been copied by most other nations.

BIRTH OF THE SAFETY BOARD

My initial employment as an Air Safety Investigator in 1966 was with the Civil Aeronautics Board (CAB). At that time, the presidentially appointed CAB Board Members basically maintained relations with the aviation communities on the national and international scenes, while busying themselves with routes, rates, and other matters concerning our commercial air carrier industry.

For all intents and purposes, the CAB Board Members left the responsibility for investigating the nation's civil aircraft accidents to personnel in their Bureau of Aviation Safety. That highly competent, technically qualified, and stable investigative staff went about their duties in a systematic, organized manner, and well deserved the esteem and rapport they enjoyed with the aviation community and general public. And we investigators never had it so good. However, in keeping with the adage, "all good things must come to an end," it was not long before that blissful balloon burst.

The descent from our zenith commenced with the Department of Transportation Act of 1968. When that came about, the Civil Aeronautics Board's Bureau of Aviation Safety became the embryo for the National Transportation Safety Board. The Safety Board consists of five presidentially appointed members who require Senate confirmation. The President designates the Safety Board's Chairman and Vice Chairman, who also require the advice and consent of the Senate.

The Safety Board was established to "promote transportation safety by conducting independent accident investigations and by formulating safety improvement recommendations." The Board writes no regulations and files no violations. Because it is nonregulatory, the Safety Board has no institutional interest to protect. Accordingly, it can be completely objective during the investigative process and when making findings and probable cause determinations.

The Board is responsible for the investigation of all civil aviation accidents, all passenger train accidents, plus major marine, pipeline, and highway accidents. The Board also conducts special studies of safety problems, evaluates other government agency transportation safety efforts, and the safeguards used in the transportation of hazardous materials. Some might question whether that is not a fairly large order for an agency with a staff of less than 400 personnel.

While there may be no pattern to transportation accidents, the Safety Board has formulated a well-designed pattern for the conduct of accident investigations. The Safety Board operates with a degree of independence that is unique on the Washington scene. The benefits derived from such autonomy are manifold, and materially contribute to its achievements. In reality, the Safety Board represents the public at the bargaining table whenever transportation safety is an issue.

The Honorable John Reed, former Governor of Maine, was the Safety Board's first Chairman. He was observed to be a perfect gentleman who exhibited many admirable qualities. He was soft spoken, easy to converse with, and exhibited a high degree of respect for the technical staff. He took a near hands-off approach insofar as the day-to-day activities of the investigative staff was concerned. Additionally, he did not initiate any policies or procedures that proved to be detrimental to the Safety Board's level of expertise or its reputation. One thing is certain, his seat was hardly cold after his term expired, before we were wishing we had him back.

President Nixon appointed Webster Todd from New Jersey to succeed Governor Reed as Chairman. During my tenure, Mr. Todd was the Safety Board's only Chairman who possessed a significant aviation background. I never had an opportunity to meet him personally; however, my cohorts in our Washington headquarters told me he was a difficult person to reason with.

He had a tendency to shoot from the hip without conducting the requisite liaison with his staff. For example, he literally opened Pandora's box insofar as the issue of parties to our investigations was concerned. This is a relatively privileged status wherein persons or organizations receiving such designations are allowed to participate in all phases of the investigations, and have access to all evidence obtained. Before his tenure, the responsibility for designating parties to Safety Board investigations was vested in the Investigator-in-Charge. After the Chairman made his edict, the Safety Board's investigators had conflicting orders and instructions. On one hand, the Chairman authorized members of a particular organization to participate; but on the other, his Bureau Chiefs denied their participation. We investigators lived with that perplexing situation for several years.

Insofar as the Safety Board's civil service employees were concerned, other adverse policies and procedures were initiated during Chairman Todd's tenure. The Executive Director, or Managing Director, (the title was changed several times) is appointed by the Chairman. Chairman Todd reassigned the reigning Managing Director, and appointed Richard Spears to fill the slot. Dick Spears was a carbon copy of that ruthless batch of rogues President Nixon brought on the Washington scene.

Before Mr. Spears arrived, the Managing Director basically had supervision of all employees, while essentially managing the Board's administrative functions and monetary policies. Mr. Charles O. (Chuck) Miller, one of the most qualified persons ever to hold the office, was the

Director of the Bureau of Aviation Safety. Chuck Miller had basic responsibility for the Board's technical functions at the time Dick Spears determined he was going to run the whole Safety Board show. The ensuing conflict between Chuck Miller and Dick Spears turned out to be a pretty heated skirmish before blossoming into royal battle on the Senate floor at a time when Safety Board personnel were over on the hill, as it's called, giving testimony.

Dick Spears probably won, but the government, the Safety Board, the aviation community, and the public, were the real losers. At any rate, Chuck Miller departed from the Safety Board scene a short time later. Not too long thereafter, Dick Spears also left the Safety Board. However, the policies his office initiated towards the technical staff have essentially remained in place. Chairman Todd resigned after Jimmy Carter was elected President.

President Carter nominated James King as Chairman, and he won Senate confirmation. Although having no aviation background, Chairman King turned out to be far easier to reason with than his predecessor. However, the managing director the new Chairman appointed proved to be an inept administrator. We referred to him as Colonel Klink, because he was bunglingly inefficient to a degree matching that of the commander of the prisoner of war camp in the popular "Hogan's Heroes" TV series. The brain drain from the Safety Board, that had begun during Chairman Todd's reign, continued under Chairman King.

After winning the 1980 election, President Reagan nominated Jim Burnett from Arkansas as Chairman. After winning Senate confirmation, Chairman Burnett appointed Peter Kissinger as the Managing Director. It soon became apparent that Chairman Burnett wanted to direct every facet of the Safety Board's operations personally, and Kissinger, exhibiting near dictatorial authority, paved the way for the Chairman to achieve his goal.

A few Safety Board employees were selected for promotion by Chairman Burnett, but he showed little regard for many others. The management style employed, and the policies and procedures initiated, caused morale throughout the Safety Board to reach a new low. The brain drain that had been slowly gathering steam turned into a virtual flood as key investigative personnel transferred to other government agencies, the private sector, or accepted early retirement.

An action causing a devastating affect on morale concerned flip-flops in two divisions. Chairman Burnett reassigned the Division Chief to the role of Deputy Chief, and promoted the Deputy to Division Chief. The law precluded him from taking such actions immediately after receiving Senate confirmation, but he made the changes as soon as permitted by law. Board employees assumed a watchful eye following those actions, fearful they might be next in line for a stab in the back.

The devastating effect Chairman Burnett's actions had on the personnel situation at the Safety Board did not go unnoticed. While it wasn't widely reported in newspapers or TV newscasts, aviation journals, periodicals, etc., were replete with articles entitled, "What's wrong with the Safety Board," or other eye-catching headlines citing the Board's problems relating to its inability to reacquire an investigative staff with the requisite qualifications.

Chairman Burnett admitted his corps of air safety investigators did not have the overall aeronautical experience of their predecessors. His claim was that they were better trained; however, it is impossible to obtain in the classroom the requisite operational background to meet the requirements of a full-fledged aircraft accident investigator. Aircraft owners, pilots, mechanics, and the government are often named defendants in huge liability lawsuits following fatal occurrences. The appointment of inexperienced investigative personnel raises questions as to whether the Safety Board was living up to its responsibilities mandated by Congress to investigate and report the pertinent facts, circumstances, and conditions related to transportation accidents.

No attempt is made to infer Chairman Burnett ever failed to act in good faith. He exhibited a high degree of loyalty and showed tireless efforts after assuming his office. The Safety Board made some significant achievements under his tutelage. He took a strong stance on some safety issues without apparent regard as to how such actions might affect him personally. However, in view of the highly technical matters involved in the Board's concentrated fields of endeavor, serious questions are raised as to whether he had the requisite background to fully comprehend the problems confronting the Safety Board.

It seems appropriate to look into the background of the Chairman who presided over the near total bureaucratization of the Safety Board. Chairman Burnett obtained his law degree at the University of Arkansas in 1973. A native of Clinton, Arkansas, he was elected municipal judge for Clinton and Van Buren County in 1974. Later in 1974, he became the juvenile judge for Van Buren County. In 1979, he became the city judge in Damascus, Arkansas. He was active in political and civic organizations. At the time of his appointment to the Safety Board, he was general counsel of the Arkansas Republican State Committee. He was a delegate to four Arkansas Republican State Conventions, and a delegate to the Republican National Convention in Detroit in 1980. He became a member of the Safety Board in December 1981 and won Senate confirmation to be the Chairman in March 1982.

President Reagan nominated Jim Burnett for another term as Chairman but the Senate failed to act on the nomination before President Bush won the 1988 election. Accordingly, in keeping with the regulations, Mr. James Kolstad, the Safety Board's Vice-Chairman, became Acting Chairman. After being inaugurated, President Bush nominated James Kolstad to be Chairman and he was confirmed by the Senate in 1990. Jim Burnett retained his membership on the Board until the summer of 1991.

Subsequent events indicated a continuation of past management philosophies. Chairman Kolstad appointed some high-level supervisors from outside sources. That provided the Safety Board with a hierarchy of dubious qualifications and unknown loyalty. Thereafter, the Safety Board underwent another reorganization. Those actions caused a further deterioration in the morale of the Safety Board's dedicated career employees.

After Chairman Kolstad's term expired, President Bush nominated Carl W. Vogt to be the Safety Board's Chairman. Mr. Vogt, an attorney with an impressive background, has been confirmed by the Senate. Besides the new chairman's other qualifications, he was a U. S. Marine Corps aviator with experience flying jet fighters off of aircraft carriers. Mr. Vogt's confirmation raises hopes that the Safety Board's downhill slide will be ending.

THE AIR SAFETY INVESTIGATOR

The Safety Board utilizes a "team" concept for its investigation of catastrophic air carrier accidents. The membership is comprised of an Investigator-in-Charge and a number of investigative specialists from its Washington, D.C. headquarters who were on the "Go Team." The Investigator-in-Charge designates the parties to the investigation. Parties normally include the FAA; aircraft manufacturer; engine manufacturer; unions representing the pilots, flight attendants, flight engineers, and mechanics; the air line company whose airplane was involved; and other persons or organizations possessing specific knowledge or expertise that might be required.

The investigation is divided into various working groups. Investigative groups are normally established for operations, air traffic control, witnesses, weather, human factors, structures, powerplants, systems, flight data recorder, cockpit voice recorder, maintenance records, aircraft performance, etc. Normally, a Safety Board investigator will chair each group. Other members of the various working groups are appointed by the designated parties to the investigation. Regular progress meetings are a part of the proceedings and a most thorough and complete investigation is conducted. In some instances, more than 100 persons will be actively involved in the methodical collection of evidence.

As evidenced by the foregoing, the Investigators-in-Charge for catastrophic air carrier accidents have technical assistance and expertise literally crawling out of the woodwork with almost no effort on their part. The same circumstances do not apply to air-taxi, business, general aviation, and less serious air carrier aircraft accidents. For those occurrences, an investigator from one of the Safety Baord's field offices is usually assigned. In many such instances, the investigators are "on their own." Upon arrival at the scene, they must make a "judgment call" determination as to the scope and magnitude of the investigation, while deciding whether technical assistants, or others with expertise in some particular field, will be needed. Accurate assessments require extensive flight experience, familiarity with the investigative process, and knowledge of a wide range of related topics.

The Safety Board has difficulty filling vacant investigative positions with fully qualified applicants. Basically, personnel with the requisite flight experience are brought on board with the expectation that they will become journeyman investigators through formal schooling and on-the-job training. The field office supervisors oversee the on-the-job training programs. The lengthy process requires the whole-hearted support of the Safety Board's journeyman investigators. Teamwork, and an aura of camaraderie, play a vital role. Selectivity is exercised by supervisors to keep from assigning newly-hired investigators to occurrences where they will be in over their heads; however, such an approach is never foolproof.

An example of the inadequate experience level of some Safety Board personnel can be gleaned from the following circumstances. A high performance, single-engine, general aviation airplane crashed while the pilot was executing an instrument landing approach. The pilot and five passengers were killed. The young, newly-hired, Investigator-in-Charge had less than 200 hours' total pilot time and was not instrument-rated. Without going into the details of the accident, it should be obvious the Investigator-in-Charge did not possess the requisite aeronautical background and degree of investigative experience required to accomplish the inquiry.

Unqualified personnel in any field of endeavor are unable to mask their lack of knowledge and experience. They unwittingly display the fact by their actions and in their conversations. This was demonstrated during the public hearing phase of the Safety Board's investigation of the Eastern Air Lines, Lockheed L-1011 accident in the Everglades Swamp near Miami, Florida. One of the nation's leading consumer advocacy groups was participating as a party to the hearing. The advocacy group's representatives made statements and asked questions that showed they lacked the extensive aeronautical background required to fully comprehend some matters under discussion.

It was exhibited again while I was attending a Boeing 747 cockpit indoctrination course in Seattle, Washington. Flightcrews from a Middle East nation were in attendance. Their captains exhibited an extensive background, but the remarks and questions by many of their copilots showed only sketchy knowledge in many areas where they should have been more conversant. In a like manner, investigators must have an extensive aviation operational background in order to know what questions to ask flightcrews, and to fully comprehend the answers given.

When an appropriately qualified Investigator-in-Charge is assigned, other parties to an investigation have much less chance of taking charge of the proceedings. Incidentially, this is a claim made with increasing frequency by pilots, operators, lawyers, and others having a vested interest in the outcome of Safety Board investigations. The situation is easily understandable. Inexperienced investigators need all the help they can get. However, they may be unable to detect when significant evidence is being overlooked, or note that the course an investigation is taking is designed to protect individuals or organizations having their own self-interests.

The journeyman investigator is not an individual who knows everything. Nobody knows everything about all airplane models. Instead of a requirement to know everything, the investigator must know when and how to go looking for help. There is nothing glamorous, romantic, mysterious, or clairvoyant about the duties of the Investigator-in-Charge. He is responsible for the time-consuming and methodical collection of evidence relating to an accident. He must be ever mindful that the possibility exists to destroy evidence that might be pertinent to the cause. In the trade, we called this breaking, or losing, the "Easter Egg." It is more prevalent when documenting the crash scene; or when examining, or otherwise disturbing the wreckage.

In many instances, it is essential that the configuration and condition of wreckage components be determined before they are moved or disassembled. Complete photographic documentation is vital to the proceedings. It is doubtful if there are any investigators who haven't missed, or possibly destroyed, significant evidence. Speaking for myself, many hours have been spent poring over photographs looking for some detail that could have been documented in a few minutes, or even a few seconds, while at the scene.

Upon notification of accidents, it is the investigators' responsibility to assure security at the crash site until they arrive. One of my investigations involved a high performance, twin-engine, turboprop airplane that crashed in South Carolina. Although of the opinion that adequate security arrangements had been made, upon arrival I found onlookers strolling through the wreckage path. When questioned, local authorities said they had been requested to furnish guard service at aviation crash sites in the past, for which payment had never been received. After being assured they would be reimbursed in this instance, guards were posted.

The airplane impacted at a relatively high velocity. Postcrash fire occurred, and the wreckage debris was scattered along a 600-foot path. A component found outside the ground fire area exhibited extensive heat distortions. The evidence showed the airplane had been subjected to both in-flight and postcrash fires. The finding was sufficient to cause suspension of the examination of the wreckage until personnel with the requisite technical qualifications and expertise could be summoned.

The airplane was equipped with a nickel-cadmium (NICAD) battery. These batteries are subject to what is known as thermal runaway. Without going deeper into the subject, it was determined the in-flight fire and subsequent crash were caused by thermal runaway of the NICAD battery.

Fire-related occurrences are probably the most difficult to investigate. The degree of difficulty is magnified when both in-flight and postcrash fire occur. Extreme caution is essential. Evidence can be destroyed by simply picking up or moving components and debris.

Investigators have to assure that they don't burn any bridges behind them. An instant replay whereby another airplane crashes in the same locale is always a possiblity. I investigated a single-engine airplane accident near a mountaintop north of Oak Ridge, Tennessee. It involved a pilot who flew straight into the mountainside while attempting visual flight in instrument weather conditions. Upon arrival in Knoxville, FAA personnel advised that an aviation enthusiast who owned a Texaco service station at the foot of the mountain would escort me to the scene.

He was very cooperative and rendered all manner of assistance during the on-scene phase of the investigation. We enjoyed a pleasant relationship and had interesting chats while driving up and down the mountain in his four-wheel-drive pickup truck. Upon our return to his station, I used a government purchase order to reimburse him, and we parted with a friendly handshake.

Exactly two weeks later, I arrived back in Knoxville, drove out to the same Texaco station, and back up the mountain in the owner's pickup truck to investigate another accident within hollering range of the previous crash. The accident involved a similar set of circumstances. The pilot flew straight into the mountainside while trying to sneak through a cloud. We all probably know "cumulonimbus" refers to a dense cloud towering to great heights. Well, for occurrences like these, investigators say the pilots flew straight into a cumulogranite cloud.

Even journeyman investigators have to learn they are always subject to Murphy's Law. Basically, that law says if it is possible for something to go wrong, it will. One of my most frustrating experiences occurred on the final investigation before retiring from the Safety Board. Everything that could go wrong, did. The wreckage was moved without proper authority. When located, the individual having custody wouldn't let me examine it. The sheriff's department personnel assigned to the case were not on duty, and it appeared all those involved or having knowledge of the accident were being uncooperative. As a result, it took about three times longer than it should have to complete the investigation of a relatively simple occurrence. Upon returning to the office, I expounded on the truth found in the saying: "It's never too late to learn."

THE AUTHOR

I was born in Richmond, Virginia in 1922, graduated from high school in Roanoke, Virginia in 1940, and enlisted in the U. S. Navy as an aviation cadet in 1942. Before graduating in August 1943, I elected to become a Marine Corps aviator, a decision I never regretted.

Knowledge about aircraft accidents began to be collected early in my flying career when I lost control of my training plane on the landing roll while still a cadet in Pensacola, Florida. Flying aficionados would say the airplane ground looped. After earning my wings, I was assigned to an F4U Corsair squadron. The Corsair was the Marine Corp's state of the art fighter with the inverted gull wing. We sometimes referred to it as the "bent wing bird." Anyway, that didn't last long because on my first flight debris was left scattered all over the runway during the landing attempt.

The Corsair was powered by a 2,000-horsepower engine with a big, three-bladed propeller. I had never flown anything so powerful. Sensing a somewhat slow airspeed late in the approach, I jammed the throttle forward and the "P" factor got me. In other words, the plane rolled over and cartwheeled down the runway because sufficient rudder and aileron inputs were not applied to counteract the propeller torque. I had three months in the hospital to think about the pilot factors involved in that mishap. Upon release from the hospital, my squadron was already overseas. The Marine Corps had notions about terminating my flying status right then and there. However, the war was not progressing too rosily in the Pacific, and possibly sensing they might need someone to blame it on if matters got any worse, they transferred me to a dive bomber squadron.

Compared to the Corsair, the Douglas, SBD, Dauntless, dive bomber was pretty docile. Its accuracy approached that of those smart weapons seen on TV during Operation Desert Storm. I accomplished nothing noteworthy while flying the SBD dive bomber in combat in the Philippines. Upon returning to the states, I completed the Instrument Flight Instructor's School and served in that capacity until discharged in 1946. In less than three months, I reenlisted in the Marine Corps as a private, was promoted to master sergeant, and ultimately got back on flying status.

I had another tour as an instrument flight instructor before joining a fighter squadron flying the "bent wing birds." Our squadron traded the Corsairs and became one of the Marine Corps' first operational jet fighter squadrons. Over the next three years, our training assignments included operational exercises off aircraft carriers before we were transferred to Korea for combat duty. Upon returning to the states in 1952, I was commissioned a Second Lieutenant for the second time, attended helicopter flight school, and returned to Korea in 1954 as a chopper pilot.

In 1956, I joined a four-engine transport squadron and gained experience flying over much of the world as the aircraft commander. In 1957, I graduated from the Aviation Safety Officer's School at the University of Southern California. Aviation Safety Officer assignments were received for most of the remainder of my Marine Corps career and those billets always provided ample opportunity to fly. Before retiring from the Corps at the end of 1965, I visited the Civil Aeronautics Board's Bureau of Aviation Safety in Washington, D. C., and made the initial contacts that culminated in employment as an Air Safety Investigator in February 1966.

My entire tenure with the Safety Board was performed in the Miami, Florida Regional Office. I investigated 679 civil aircraft accidents involving gyrocopters, balloons, helicopters, and fixed wing models from ultralights, gliders, homebuilts, and the tiny Piper Cub, to the Boeing 747. The experience gained while flying a variety of aircraft in the Marine Corps was very beneficial.

In 1985, I retired after a 19-year tenure with the Safety Board. Two and one-half years later, I was reemployed for the specific purpose of completing the accident reports for investigations initiated by personnel who had departed the Safety Board scene, or those still on board with excessive backlogs. At the end of 1988, I permanently retired from the Safety Board.

In some chapters, lighthearted comments have been made about Safety Board doings. In others, forthright expressions are made concerning my opinions of the detrimental effects produced by the management philosophies of some of the presidentially appointed Safety Board Chairmen. However, I am now, and will continue to be, a strong Safety Board booster. Our country needs, and must always have, a totally independent and unbiased organization, manned by a staff possessing the requisite expertise to conduct in-depth investigations of its catastrophic transportation disasters. Accordingly, my zeal for the Safety Board parallels my allegiance to the U. S. Marine Corps.

THE MOTIVATOR

It is believed that any creative endeavor requires much self-discipline. Just getting started can pose a massive problem. Insofar as this book is concerned, I fell victim to procrastination until visiting a museum in Santa Fe, New Mexico, where I read the following statement by Jesse L. Nusbaum, an early photographer:

> **"THE CREATIVE INSTINCT LIES DORMANT IN MANY OF US AND ONLY WAITS FOR THE PROPER CONDITIONS TO MANIFEST ITSELF"**

Jesse prompted me to commence this project. And, a copy of his statement, still carried in my wallet, jolted me out of my lethargy on numerous occasions when I became distracted by other activities.

\- -

I always wanted to emulate my mother, Epsie Baldwin Watson Vale. Even while rearing her 11 children and my oldest sister's four offspring after both of their parents early demise, Mom found time to bring comfort to those in need while teaching the good news about Christ's coming. Then after all the children had gone their separate ways, she dedicated the remainder of her life to helping others regardless of their race or creed. An apt description of her activities would be that she became a One-Woman Salvation Army in the town of Farmville, Virginia. All these deeds were accomplished while possessing only meager monetary resources. She was selected Virginia's Mother of the Year in 1967 at age 69. Thereafter, she continued her good works for another 18 years when failing health forced a slowdown.

But early-on it became evident that I could not be like my mother. I am too emotional. When attempting to comfort others I get all choked up. And my efforts at one-on-one evangelism were always pathetic. Then I remembered from Christ's parable teachings that the talents were not distributed equally among the servants. One was given five, another two, and the other one. Each servant was given according to his ability. However, when I began searching for whatever talents I had to use or give, I still came up short. Since I never made a wise investment, monetary resources were not available to accomplish anything significant in that arena. There had been instances during my Marine Corps career and employment by the National Transportation Safety Board when some favorable comments were heard concerning my writing style. So I turned to God and asked that he be my partner in this book writing endeavor. Depending on our degree of success, some who are hungry will be fed, others that are homeless will be housed, and hopefully, many will be brought to Christ.

To aid in our efforts I joined literary clubs, attended writer's conferences, and read almost everything on the local library's shelf about the book publishing business. These activities painted a very gloomy picture. It was learned most publishers don't like to take chances with unknown authors. When they do decide to accept an unknown's work, the business arrangements invariably favor the publisher.

During initial attendance at yet another literary club, J. Patrick O'Connell, a former publisher with a treasure of knowledge, was one of the principal speakers. Pat recounted some of his experiences, mentioned the difficulties unknown authors could encounter with publishers, and suggested that there were occasions when self-publishing might be a viable alternative.

I pondered Pat's comments for two weeks before giving him a call. After identifying myself, I gave him a brief description of the subject matter of the book, and asked point-blank if he would act as a consultant on a publishing venture. Some hesitation was perceptible before Pat replied that it was an interesting proposal; however, he would have to read the manuscript before giving a definitive answer.

I experienced two weeks of anxiety and hopeful anticipation after delivering the manuscript to Pat. When we met again, Pat handed me his analysis that included the following:

The book is well written but needs suggestive and grammatical editing. Lots of potential and with luck could fly high.

After coming to an agreement on the business arrangements, we set about our tasks. Pat became the president of Harbor City Press after it was incorporated in Florida. All the while, the manuscript was being polished, edited and revised. Pat performed admirably in every respect. The experts he brought on the scene proved to be seasoned professionals. I followed his recommendations almost carte blanche. On those occasions when we arrived at different opinions, we ironed out our disagreements in an amiable fashion.

I came to realize Pat's expertise was essential long before this book went to the printers. While not knowing whether it will be accepted by the reading public as this is being written, the prospects for UNHAPPY LANDINGS were greatly enhanced by Patrick's input and suggestions. These happenings, in conjunction with the timely turn of many other events, made me believe God was responding to my request for assistance.

We appreciate your attention and your perseverance in reading our finished product. Godspeed in all your endeavors, and a happy arrival after every trip regardless of the mode of transportation you may use.

INDEX

Each chapter in this book "stands alone." The index has been broken down by chapters.

CHAPTER ONE - MISLEADING INFORMATION

CHAPTER TWO - THE INFALLIBLE BREED

CHAPTER THREE - HAULED INTO COURT

CHAPTER FOUR - FACE TO FACE WITH THE ENEMY

CHAPTER FIVE - WIND SHEAR, A DEADLY FORCE

CHAPTER SIX - PEOPLE AND PLACES

CHAPTER SEVEN - WITNESS FOR THE DEFENSE

CHAPTER EIGHT - OUR MAN IN SOUTH CAROLINA

CHAPTER NINE - TOOLS OF THE TRADE

CHAPTER TEN - WEIGHT WATCHING

CHAPTER ELEVEN - IN SEARCH OF PERFECTION

CHAPTER TWELVE - ON BETTERING OUR WAYS

CHAPTER THIRTEEN - AFRICAN ADVENTURE

CHAPTER FOURTEEN - THOSE DULL PUBLIC HEARINGS

CHAPTER FIFTEEN - JUST THE FACTS, MA'AM

CHAPTER SIXTEEN - LIKE AN OAK LEAF IN A TWISTER

CHAPTER SEVENTEEN - DETECTIVE STORIES

CHAPTER EIGHTEEN - THE ACCIDENT THAT NEVER WAS?

CHAPTER NINETEEN - MEDICAL MEANDERINGS

CHAPTER TWENTY - HUP-TWO-THREE-FOUR

CHAPTER TWENTY ONE - BEWILDERED

CHAPTER TWENTY TWO - AMERICA'S KAMIKAZE CORPS

CHAPTER TWENTY THREE - FLAT OUT DANGEROUS

CHAPTER TWENTY FOUR - THE GREMLINS DID IT

CHAPTER TWENTY FIVE - POTPOURRI

EPILOGUE

APPENDIX ONE - AN INVESTIGATION WALK-THROUGH

APPENDIX TWO - BIRTH OF THE SAFETY BOARD

APPENDIX THREE - THE AIR SAFETY INVESTIGATOR

APPENDIX FOUR - THE AUTHOR

UNHAPPY LANDINGS is available at quantity discounts for use by or for -

Flight schools, aero clubs, pilot groups, other aviation oriented organizations, and like entities promoting aviation safety -

Charitable fund-raising organizations -

The promotion of products or services -

Other legitimate endeavors -

For information write or call:

Harbor City Press, Inc.

P. O. Box 361033

Melbourne, Florida 32936-1033

Phone: (407) 255-0908

- -

TELEPHONE SERVICE FOR VISA
AND MASTERCARD CUSTOMERS

UNHAPPY LANDINGS may be ordered by telephone. All telephone orders must be charged to either your VISA card or MasterCard. The price of $18.95 includes $2.00 for shipping and handling. Please initially dial 1-800-247-6553 for telephone orders. If any difficulty is experienced, kindly place your order by dialing either of the other numbers listed below.

1-800-247-6553

1-800-741-4890

(407) 255-0908

- -

NOTES

GIFT SHOPPING MADE EASY

NOTES